Performance for Resilience

"An important guide for creatively engaging youth on the most critical topic our planet faces today—climate change. Youth not only have a right to be a part of building a resilient future in the face of climate change, they also have the passion and creativity to advance all of our thinking. This book provides practical tools and inspirational stories for involving youth in conceiving of a better tomorrow through the arts."

—Dr. Victoria Derr, *California State University Monterey Bay, USA*

"Through co-created climate communication, Osnes persuasively advocates for the centrality of creative youth participation in building climate resilient communities. In sharing her experiences of using collaborative creative metods, she facilitates an important and vital shift for climate communication and engagement strategies, towards the inclusion of young people's embodied and place-based knowledge as central to achieving inclusive climate resilience."

—Professor Julie Doyle, *University of Brighton, UK*

"This is a project that rouses a sleeping giant—youth—to joyfully engage in authoring a new story for climate, energy, and resilience. It supports them in using their voice through art to act on the scientific knowledge that our climate is changing. This is an educational tool well-suited to anyone who trusts that youth need to be a part of our collective solution moving forward."

—James Balog, *Director, Extreme Ice Survey and Earth Vision Institute*

Beth Osnes

Performance for Resilience

Engaging Youth on Energy and Climate through Music, Movement, and Theatre

Beth Osnes
University of Colorado Boulder
Boulder, CO, USA

ISBN 978-3-319-67288-5   ISBN 978-3-319-67289-2   (eBook)
DOI 10.1007/978-3-319-67289-2

Library of Congress Control Number: 2017952820

Cover illustration: Pattern adapted from an Indian cotton print produced in the 19th century

Printed on acid-free paper

This Palgrave Macmillan imprint is published by Springer Nature
The registered company is Springer International Publishing AG
The registered company address is: Gewerbestrasse 11, 6330 Cham, Switzerland

This book is dedicated to Gerald Vandever, dedicated teacher and ally to indigenous youth. His love of youth and performance gave this project its first wings and set it on its course.

PREFACE

ABOUT THIS BOOK

When I was in fourth grade, an environmentally concerned teacher taught my class this song, *"Pollution, pollution, you can use the newest toothpaste, then rinse your mouth with industrial waste!"* It was funny and catchy and 43 years later, I still remember the words, the tune, and the lesson (a miracle, given that every other lesson has long since faded from memory.) As an adult, this strategy of using theatre to engage has been evident in all my work. Recently, I found myself simultaneously working on projects that either used performance as a tool to educate and engage, or for women to empower their voices, or to effectively communicate about climate change. Eventually, these disparate interests and projects naturally wove together into one project that engages youth in speaking and acting for a resilient future. This project is *Shine*.

This book shares the story of *Shine*; it was written primarily for educators of youth who seek active methods for engagement in issues related to energy, climate, and resilience. Within these pages are ideas for youth engagement, ranging from hour-long modules, to a multi-week mounting of a production. Readers of this book will be able to access the script, music, and choreography for *Shine* via its accompanying website http://www.insidethegreenhouse.org/shine in order to mount a production with local youth in their school or community. Additional materials include: warm up exercises; materials for creating curricula; interviews with collaborators to share multiple perspectives;

and a detailed description of how the performance was mounted at each location on its international tour—complete with lessons learned, recommendations, and feedback from participants and audience members. There are multiple levels of possible engagement with *Shine*. If time and resources are limited, educators and community organizers may want to show their students the video of the full performance of *Shine* (http://www.insidethegreenhouse.org/shine/), and merely engage youth in conversation about issues introduced in the show—utilizing the discussion questions provided in the Chap. 2. If more time is available, Act One of the full performance can be shown, and teachers can use the Act Two module to guide students in creating skits for their own Act Two, finding solutions to climate, energy, and resilience challenges that they identify from their own community. Alternately, any modules described in Chap. 2 can be used and contextualized by showing or assigning the video of the full production. Various parts of the performance can be used independently to address specific areas of study or concern. For example, an educator may want to engage students in staging the *Harvest* module to actively explore differences between urban and rural lifestyles and values in regard to energy. The performance materials, the lessons learned, and the recommendations within this book contribute towards a greater understanding of the use of performance as a tool for activating youth in authoring, rehearsing, and sharing a vision for the future they want. As seen in the description of Malope Primary School in South Africa in Chap. 3, the resilience focus of your community may be different from the focus of *Shine*. In that case, you may want to use *Shine* as an example for how performance can be a useful tool for exploring resilience issues vital to your community and create a new performance piece entirely.

Material in this book and its accompanying website can easily be used towards a standards-based curriculum. This project provides materials for the easy creation of a curriculum module that can be utilized in the USA to reach Common Core standards by teachers of English language arts/literacy (ELA) or for Next Generation Science Standards (NGSS) or state science standards. Similar academic standards in other countries could likewise be met by creating modules from this material. The learning goals for youth participants of *Shine* include:

Subject Area Learning Goals:

- understanding the relationship between energy and climate (Science)
- placing the earth's production of fossil fuels and the impact of human-use of fossil fuels into scale within the last 300,000,000 years of geological history (Geology)
- understanding how our energy-use is impacting climate (Science)
- placing all this within the context of human history and society (Sociology, History, Civics)
- understanding effective climate communication by critically engaging with the script for Act One, and by inviting youth to author Act Two, dramatizing local solutions towards a positive climate future (English, Media, Science, Civics)

Participatory Performance Learning Goals:

- embodied learning
- nuanced understanding of themes through physical participation in dramatic metaphors
- youth empowerment
- civic engagement
- collaboration with others to effectively communicate youth-authored solutions to an audience
- students as authors of knowledge and partners in public display

The over-arching goal of this project is to create a performance experience for youth that guides them through an embodied exploration of the climate legacy they have inherited. *Shine* further works to inspire and equip youth to create local solutions for inclusion in their city's plan for resilience. One of the objectives for achieving this goal is to have youth immersed in artistic excellence and the pre-eminent science on climate and energy, through the process of rehearsing Act One, authoring Act Two, and performing the show before a larger audience. Having this project professionally created, field-tested, and published as a book is intended to lend validation to the use of performance as a tool for youth engagement in city planning.

Educators and community organizers may wish to use the example of *Shine* to inform and inspire the creation of their own performance experience for community engagement. Researchers will find a useful case study of participatory culture for activating citizens in imagining and realizing a more sustainable future. Due to the interdisciplinary nature of this project, this book will be of interest to undergraduate and graduate students of many disciplines, including: Theatre, Performance Studies, Environmental Studies, Environmental Design, Communication, and Geography. Educators from K-12 can approach the use of this book as a module they can use in constructing their educational plan. Given the expansive scope of the performance, it can advance students in the study of: climate, energy, civics, human development, history, composition, music, and theatre. City planners and organizations working to support inclusive city planning can also gain a valuable model through this participatory tool for community engagement.

How This Book is Organized

Chapter 1 introduces *Shine*, providing the background for how and why it was created, its theoretical foundation, and a synopsis of the story it dramatizes. Chapter 2 includes the script, as well as interviews with the composer, the choreographer, and one of the primary partnering scientists. Materials for creating curriculum for facilitating youth in mounting this performance is provided for possible adaption for K-12 schools, youth organizations, faith organizations, or youth camps. There are also instructions for various performance exercises and games to do with youth participants to prepare them for being expressive. Chapter 3 shares the outcomes from the international tour of *Shine*. Detailed descriptions of each mounting within every hosting community reveal the many ways the performance can engage students. Chapter 4 culminates in a concluding summary of lessons learned and recommendations from the tour, including useful approaches, strategies, and best practices. Theories are articulated on the use of performance for youth-sparked community engagement for climate, energy, and resilience planning. Recommendations for the further use of performance for resilience planning are also provided.

Personal Motivations

My fourth-grade schooling in the potency of performance and engagement has been affirmed in my personal and professional work ever since. I've witnessed how speaking out and physically acting embodies our ideas and makes it more possible and probable to enact change in our lives. *Shine* is the culmination of everything I've learned. I'm committed to an embodied approach, so as to draw from the ingenuity of the body and all it knows and can do. I'm committed to an art-based approach, so as to draw on the power of the dramatic metaphor to tease out a more nuanced understanding of both problems and solutions. I'm committed to collaborating with wonderful scientists (those with a sense of wonder) who value engaging the power of youth to put their good work into service for the planning of a bright tomorrow. And lastly, I'm committed to an approach that is fun, because those things that bring us joy are ultimately most sustainable. We will continually come back to something again and again if it makes us happy.

Boulder, USA   Beth Osnes

ACKNOWLEDGEMENTS

I extend my gratitude to the University of Colorado's Environment and Sustainability Seed Grant that provided support for the development of this project and for its international tour. I thank my collaborators: Tom Wasinger, who composed the music; Arthur Fredric and Lisa Denning, who developed the choreography and provided direction for the New York and Connecticut performances; Meridith Richter who edited the footage and created the website for *Shine*; and my primary scientific collaborators, Paty Romero Lankao with National Center for Atmospheric Research (NCAR), Joshua Sperling with the National Renewable Energy Laboratory (NREL), and James White, Professor of Geological Sciences and Director of the Institute of Arctic and Alpine Research at CU. I extend my thanks to CU Physics prof. Noah Finkelstein, who is also director for the Center for STEM Learning, for guiding the curricular efforts associated with this project. My co-directors of "Inside the Greenhouse," Max Boykoff and Rebecca Safran, provided invaluable support for this effort in so many ways. I thank my research assistant, Shira Dickler, for her contribution to the materials for the curriculum within this book. I thank my editors at Palgrave Macmillan for their assistance and support, and my copy editor Juliana Forbes for her ever-watchful eye and wisdom with words. My family is ever supportive of my work. Thanks to them for traveling with me to many parts of the USA and the world to engage youth in this performance. My daughter, Lerato, has performed alongside youth in many communities; my husband, JP, has served as our designer and production director on several occasions;

my son, Peter, was a stagehand for the NCAR show; and my daughter Melisande provided graphic design and co-facilitation in South Africa. I also extend my gratitude to the many community hosts for *Shine*, who are all mentioned by name in Chap. 3. Most of all, I thank the youth throughout the world who shine a light on our sustainable future with their spirited participation.

CONTENTS

1 Introduction to *Shine*   1
Background   4
Theoretical Foundation   5
Synopsis of Shine   7
Notes   13
References   15

2 Design of *Shine* **as a Method for Engagement**   17
Script with Stage Directions and Sound Cues   19
Notes from the Composer   31
Notes from the Choreographer   34
Interview with Partnering Energy Engineer   37
Materials for Building a Curriculum   39
Additional Discussion Questions   56
Sample Exercises and Games   57
Notes   60
References   60

3 Outcomes from the International Tour of *Shine*   61
Tuba City, Arizona, Navajo Nation   63
Boulder, Colorado, USA   69
New York City, New York   88
London, Great Britain   93

New Orleans, Louisiana 104
Malope, South Africa 114
Connecticut, USA 122
Boulder, Colorado 125
Notes 131
References 133

4 Conclusion 135
Contributing to Youth 136
Youth Contributions 139
Performance 141
Conclusion 146
Notes 148
References 149

Appendix 151

Index 155

LIST OF FIGURES

Fig. 1.1 Paty Romero Lanko (left) with participates of *Shine*
in rehearsal   2
Fig. 1.2 Ancient plants   8
Fig. 1.3 Ancient plants and animals get covered by mud   9
Fig. 1.4 First fire by humanity with carbon being released   10
Fig. 1.5 Harvest dance   11
Fig. 1.6 Youth performers enact skit for Act Two of *Shine*   13
Fig. 2.1 Students playing a game before rehearsing *Shine*   57
Fig. 3.1 Tuba City High School students and the Hopi
Language Arts teacher dancing in the final number *Shine*   64
Fig. 3.2 Weaving the fabric of community   70
Fig. 3.3 Rehearsing the Machine exercise   71
Fig. 3.4 Skit dramatizing the growing of a tree   83
Fig. 3.5 Fossil fuel flags   89
Fig. 3.6 Foss and Sol in a bout of sibling rivalry   93
Fig. 3.7 Students at Saint Dominic's School responding to the
question "Who wants to be the trilobite?"   105
Fig. 3.8 A male student portraying the mother explains
menstruation to a female student portraying the
daughter   115

Fig. 3.9   Foss and his followers dancing   123
Fig. 3.10   Geological timeline   126
Fig. 3.11   Enacting a scene about a local store where the currency
              is green carbon credits for Act Two   128
Fig. 4.1   The release of balloons after the final song in *Shine*   137
Fig. 4.2   Adults play in the process of rehearsal for *Shine*   141

CHAPTER 1

Introduction to *Shine*

Abstract This chapter introduces *Shine*, a performance experience for youth that is designed to engage participants in resilience planning. It weaves climate science and artistic expression into a story that spans 300 million years of geological time to convey how energy, humanity, and climate are interrelated. Through humor, music, and movement, *Shine* physically engages youth and leads participants to embody aspects of climate science and human development that ultimately led to this current moment—where our use of fossil fuels is impacting our climate. This chapter provides the background in the Rockefeller Foundation 100 Resilient Cities Initiative, the theoretical foundation for this examination of *Shine*, and a synopsis of the story.

Keywords 100 Resilient Cities · Youth engagement · Applied theatre Climate change · Energy · Resilience · Ecodramaturgy

Shine is a performance experience for youth that is designed to engage participants in issues of climate, energy, and resilience (http://www.insidethe-greenhouse.org/shine). It weaves climate science and artistic expression into a story that spans 300 million years of geological time to convey how energy, humanity, and climate are interrelated. Through humor, music and movement, *Shine* demystifies scientific facts and concepts, and physically engages youth in what are usually cerebral abstractions. Rehearsing the musical immerses youth in the lexicon surrounding climate and energy and leads participants in embodying aspects of climate science and human

© The Author(s) 2017
B. Osnes, *Performance for Resilience*,
DOI 10.1007/978-3-319-67289-2_1

1

development that ultimately led to this current moment—where our use of fossil fuels is impacting our climate. The first half of the show is professionally scripted, composed, and choreographed to convey the story that has already been told by history. The second half—our future story—is authored by local youth to generate solutions for their city's energy, climate, and social challenges. The solutions they conceive in Act Two are legitimate responses to future challenges and can effectively build resilience for their community. Local youth are led in performing the show in each location. The performance experience of *Shine* is designed to support youth engagement in community resilience planning.

In each city where this show has been mounted, local stakeholders have served as hosts of the performance experience and champions of the effort. When the show was mounted in Boulder Colorado, one such host was Paty Romero Lankao, an interdisciplinary sociologist working as a senior research scientist at the National Center for Atmospheric Research (NCAR), leading the "Urban Futures" initiative. She participated with *Shine* from an impressively well-informed perspective, being a co-leading author to Working Group II of the Nobel prize-winning Intergovernmental Panel on Climate Change (IPCC) Fourth Assessment Report (AR4) and convening author of IPCC: AR5, North American chapter (Fig. 1.1).

Fig. 1.1 Paty Romero Lanko (left) with participates of *Shine* in rehearsal. Photo by Conner Callahan

The morning of the rehearsal, Lankao hosted the entire cast of local youth in the finest conference room NCAR boasts—where IPCC negations took place and where the dramatic start of the Rocky Mountains fills the floor-to-ceiling windows. In preparation for 13 young performers, she pushed tables to the side to make room for an all-day rehearsal, with snacks of grapes, crackers, and cheese sticks and costumes strewn around the carpet. Through Lankao's position at NCAR, she claimed this place of positive social power for the expression of youth voices authoring 'urban futures.' Throughout the day, she rehearsed the dances and movements alongside the youth, sometimes elucidating a scientific principle or idea referenced in the script. With a group of four young people, she created and performed in a scene for Act Two that focused on the importance of maintaining our forests to avoid global warming. In the late afternoon, she performed in our public showing of *Shine* for NCAR scientists, invited guests, and the general public. No one present could miss the crystal-clear messages she sent through her example: the contributions of young people matter; and performance is a valid method for community engagement in authoring our city's future.

As an Associate Professor of Theatre and Environmental Studies at the University of Colorado at Boulder (CU), I wrote and created this performance experience in collaboration with nationally recognized performing artists and climate scientists. Three-time Grammy winner, Tom Wasinger composed the music, and Arthur Fredric, master teacher with the New York City National Dance Institute and former Broadway performer, developed the choreography. Primary scientific collaborators include: Lankao with NCAR; Joshua Sperling with the National Renewable Energy Laboratory (NREL); and James White, Professor of Geological Sciences and Director of the Institute of Arctic and Alpine Research at CU. CU's Environment and Sustainability Seed Grant provided support for the development of this project and for its international tour. I travelled to each location of the tour to facilitate the performances with local collaborating host institutions. The intention of the tour was to learn best practices from each city's process to contribute to a deeper understanding of how performance can effectively engage youth in authoring their city's plan for resilience. To date, *Shine* has been performed by local youth in seven different communities, four of which are cities that are part of the Rockefeller Foundation 100 Resilient Cities Initiative: Boulder, CO, USA (June 2015, October 2015, and July 2016), New York City, NY, USA (October 2015), London, UK (January 2016), New

Orleans, LA, USA (April 2016), and three additional cities: Tuba City, Arizona within the Navajo Nation (March 2015), Malope, South Africa (June 2016), and Brookfield, CT, USA (July 2016).

BACKGROUND

The Rockefeller Foundation launched the 100 Resilient Cities Initiative to help cities around the world become more resilient to the physical, social, and economic challenges that are a growing part of the twenty-first century.[1] Boulder, Colorado was among the first group of 32 cities chosen by this initiative in December of 2013. I was invited to attend Boulder's first all-day meeting with community stakeholders and the Rockefeller Foundation 100 Resilient Cities team in May of 2014. There I saw an opportunity to contribute using performance and youth engagement in authoring our city's plan for resilience. I was attracted to working with this initiative for three reasons: (1) it would bring my community-performance work into the cities throughout the world that are leading the movement in resilience planning; (2) it provided an international platform for sharing best practices; and (3) it had an inclusive approach towards resilience. As stated on their website, "100 Resilient Cities supports the adoption and incorporation of a view of resilience that includes not just the shocks—earthquakes, fires, floods, etc.—but also the stresses that weaken the fabric of a city on a day to day or cyclical basis."[2] My definition of resilience—refined and utilized throughout the process of creating, rehearsing, and touring this performance—is a compilation of definitions largely provided in a publication by the Institute for Social and Environmental Transition, *Beyond Resilience*.[3] Resilience is the capacity of our communities to function, so that everyone—particularly those who are under-resourced and vulnerable—survive and thrive, no matter what social stresses or climate shocks come our way.

My interest in cities arose in response to the growing international focus on cities as an appropriately scaled measure of community belonging. It seems that the feeling of belonging within a city has the potential to unite a community amid differences perhaps more successfully than our often-divided sense of nationalism. As we move into the future, it is believed by many that "cities will be where most decisions and actions occur that impact climatic, geophysical, atmospheric and ecological processes that define critical 'hard-wire' thresholds in the Earth's environment."[4]

This performance experience was created through "Inside the Greenhouse" (ITG), an endowed initiative inspiring creative climate communication that I co-founded and co-direct at CU with Rebecca Safran, Associate Professor of Evolutionary Biology, and Max Boykoff, Associate Professor of Environmental Studies. For four years, I have co-taught a course on creative climate communication and have co-planned events and student internships through ITG. My creation of *Shine* is in large part my attempt to put best practices learned from being a part of ITG into play, literally and figuratively.

THEORETICAL FOUNDATION

Shine is a performance that was created to dramatize the tensions inherent in energy use that contribute significantly to climate change. It is designed to serve as an activating framework that can encompass multiple narratives, allowing authorship by local communities that reflect each city's unique challenges and solutions. The goal is for young citizens of cities to be facilitated in coming together as a community to actively participate in authoring a more resilient plan for their community. The design for this performance experience is based on the belief that if people are guided in proposing solutions aligned with their values and priorities, they are more likely to feel ownership for and act on those solutions.[5] Instead of top-down, expert-driven dissemination of information this project seeks to harvest local knowledge and invigorate community-based solutions. Embodied and fun, this performance is designed to be a fresh way of inviting a wider constituency into the planning process for community resilience. There is evidence that embodying concepts in this way is beneficial to learners.[6] This project is based on the principal of active culture that reflects the recognition that people frequently get more out of making art than seeing the fruits of other people's labors.[7] If citizens participate in the creation of a performance, then the process will likely help guide their thinking in, and most importantly, out of the theatre.[8] Focus on local resilience issues brings both social and climate solutions to the community level. Many people feel overwhelmed by climate change when it is framed as a global crisis and disengage from the issue. Research on climate communication reveals that the most effective scale for framing climate change is at the local level—especially with an intent to engage.[9]

One way of categorizing this performance is within Community-Based Adaptation to Climate Change, which is a relatively new field that focuses on innovative participatory methods that are being developed to help communities: analyze the causes and effects of climate change; integrate scientific and community knowledge of climate change; and plan adaptation measures.[10] This is closely aligned with an emerging practice called "ecodramaturgy," a term coined by Theresa J. May, one of the editors of *Readings in Performance and Ecology*, which she defines as "theater and performance making that puts ecological reciprocity and community at the center of its theatrical and thematic intent. Ecodramaturgy carries with it new frames for thinking about theater and new approaches and challenges to making theater."[11] This approach is supported by ecocriticism, the interdisciplinary study of the relationship between literature and the physical environment which takes an earth-centered approach to literary studies.[12] One of the basic propositions of ecocriticism is that "environmental problems require analysis in cultural as well as scientific terms, because they are the outcome of an interaction between ecological knowledge of nature and its cultural inflection."[13]

My approach is also informed by Applied Theatre scholarship and practice, which enlists the participation of non-performers, often in non-traditional performance spaces, and uses performance as a tool to work through areas of concern that participants identify.[14] In this work, what is gained through the process is valued as highly as any resulting product. Also influential to this project and often included under the umbrella phrase "Applied Theatre" is Theatre in Education, described by practitioner David Pammenter as, "an art form that examines, questions and represents the realities of our current human condition and makes new meaning in pursuit of progressive change and positive human development."[15] The field of Performance Studies also offers a useful lens for viewing this project, especially in the way Performance Studies pioneer, Richard Schechner, grounds the conception of performance as, "an elusive, playful, embodied, multifaceted, protean operation."[16] Schechner also points out that, "Those practicing performance studies are actively involved in community life, often becoming advocates for, or coactivists with, those they are studying."[17] Borrowing from the language of Shannon Jackson in *Social Works: Performing Art, Supporting Publics*, this performance could be considered a "social work": "socially-engaged, cross-disciplinary performance projects in which social goals and aesthetic forms have a kind of structural coincidence."[18] Jackson goes on to comment how the artistic structure or framing of a performance project can interact with its community engagement aims. She argues that, "To imagine an infrastructural aesthetic is not only to take a community stance on the arts but also to take an aesthetic stance on community engagement; it asks what the aesthetic frame does to and with the idea of community and what the aesthetic process does to and with social processes."[19] This project is also inspired by conceptual idealists in city planning, such as Tony Fry, who recognize that, "existing approaches to the challenges are insufficient in the face of what climate change and global unsettlement will confront humanity with in the coming decades and centuries," and that what is needed is, "another conceptual framework to explore, think and create modes of worldly occupation about how to more adequately deal with the future that is coming towards us."[20]

At the heart of this project are the expressed beliefs, creations, and ideas of the youth of each city. Research shows that by involving youth, especially from underprivileged communities, in planning and implementing urban improvements, there are enormous benefits for youth, for the wider society, and for the future.[21] Actively involving adolescents while they are still relatively young is important regarding climate-related issues, since research reveals that pessimism about addressing climate change increases with age—particularly from early to late adolescence.[22] This performance project recognizes youth as potential thought-leaders and idealists, not just as future adults but as current members of their communities. Youth often constitute a sector of heavy users of the city's infrastructure and resources. Through *Shine*, performance is being used to diversify who has a voice in a city's planning and whose needs are prioritized, such that the city benefits from the contributions of every sector, especially youth, who are traditionally under-represented yet are, arguably, any city's greatest asset.

Synopsis of *Shine*

The play opens about 300 million years ago with the Sun noticing Earth's ancient plants and animals doing a musical number, yes, a musical number (Fig. 1.2). Life during the Carboniferous Period is getting more and more animated. Green leggy plants—with leaves draped around them—grow and stretch their way through the audience and onto the stage, gathering around the Sun. Lurking behind are the ancient creatures—dragonflies, scorpions, and trilobites—licking their

Fig. 1.2   Ancient plants. Photo by © 2016 Steven Sutton, DUOMO

proverbial lips at the plants. As the Sun deciphers the plants' choreography more closely, she realizes they are storing up her energy by dancing the various parts of photosynthesis. Simply put, they gather energy from the Sun, take up water through their roots—which excites their cells—breathe in CO_2 from the atmosphere, and grow. The animals eat the plants. They all eventually die and get covered up by layers of mud, rocks, and sand (Fig. 1.3).

This cycle goes on for millions of years. From beneath the Earth emerges a guy named Fossil Fuels—who prefers a slightly flashier name, Foss. He bumps into the Sun and claims to be her little brother. She protests, confidently stating the scientific fact that energy can neither be created nor destroyed, so she can't have a *new* energy brother. He retorts that energy can't be created or destroyed but that it can be transformed from one form into another. He explains how he is the result of this repeated cycle of her energy getting stored up and compressed beneath the surface of the Earth, and that since he is another form of her energy, is rightly a next of kin. She begrudgingly accepts him as family and continues to shine on her planets for a few million more years. Foss slumps

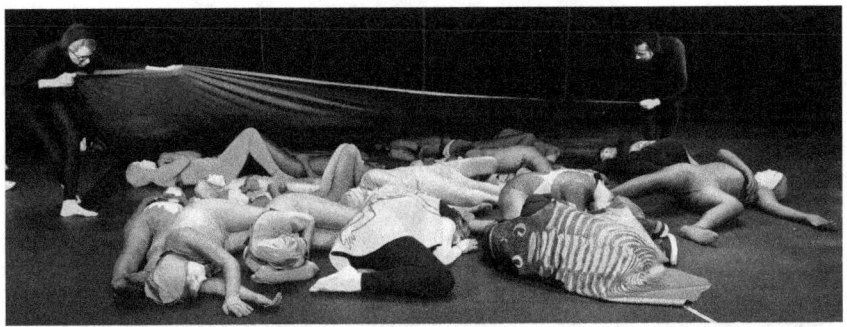

Fig. 1.3 Ancient plants and animals get covered by mud. Photo by © 2016 Steven Sutton, DUOMO

down next to her with a mix of boredom and impatience that is interrupted during the Triassic Period by their sighting of a dinosaur. Both the Sun and Foss are surprised and delighted with this new lifeform. Foss watches into the distance as it proceeds towards extinction. He slumps down by his big sister, who has resumed shining on her planets. Time fast-forwards to 2.8 million years ago, and Foss spots the first humans entering the stage. Marveling over this new life form, the siblings watch the humans rub sticks together until a red flame licks upward. The Sun marvels at these creatures' cleverness in using her energy—stored in biomass—to warm themselves. Foss asks what the black stuff is that is being released from the fire, and Sol reassures him that it is just carbon being released into the atmosphere, but that it is no big deal since it is such a small amount (Fig. 1.4).

When lead-in music for the next dance interrupts, Sol excitedly invites her brother to join the humans in their harvest dance, noting that sometimes life on this planet does musical numbers. Too cool, he opts to sit this one out and exits. The humans enter in procession—each with a roll of recycled newspaper cut into strips at one end—and mime the actions of spreading seeds, tiling, and growing the food that sustains their lives. In turn, each farmer kneels and pulls the inner-most strips of paper from within the roll to mimic the growth of the plant (Fig. 1.5). Eventually, Foss and his troupe can't resist crashing through the harvest line—in doing so, knocking over the plants and disrupting the harvesters—with an upbeat number glorifying a fossil fuel-based, urban lifestyle. The

Fig. 1.4 First fire by humanity with carbon being released. Photo by © 2016 Steven Sutton, DUOMO

resulting mash-up of their two tunes and choreography bonds the two siblings and the communities into a lively synthesis of the urban/rural lifestyles. For the time being, they figure out a way to live together in harmony and with vitality.

Meanwhile, the harvesters begin to settle into a city and create a human loom to literally weave together decorated paper strands—representing individuals—into a community fabric. Foss's impatience with their solemn pace spurs his discovery of what could be his purpose: to help humanity move faster and with more power. Sol warns her brother that humans may not be able to handle him, what with millions of years' worth of solar power packed into him. Foss argues for the benefits of fossil-fuel-based progress that could relieve the toil of human struggle. Sol argues for the benefits of a sustainable balance with nature, just using solar and biomass energy. Their parting argument culminates with her

Fig. 1.5 Harvest dance. Photo by © 2016 Steven Sutton, DUOMO

stern warning that somebody could get hurt and his taunt that she just might be jealous of her brother's potential impact.

The harvesters—living a sustainable lifestyle—proceed to form a human loom: a square with four people at each side. They perform a pastoral movement piece to music composed from the sounds of actual wooden looms. One side of the human loom walks towards the opposite side, hands the opposite line the end to their scrolls, backs up to unfurl each, revealing the colorful strips that are decorated with identifying aspects of their community. Once this woven masterpiece is revealed, Foss re-enters with a flag-bearing army behind him leading the way towards progress fueled on by a lurching industrial rhythm. They circle the fabric of community as Foss sings lyrics beckoning others to help him fuel the world with coal, oil, and gas. Decorated flags depicting all the ways in which the hosting community is using fossil fuels, are waved as the group executes their synchronized mechanical steps. Black carbon is released in enormous amounts, which are symbolized by small bits of black tissue paper tossed by Foss. Triumphantly Foss boasts to his sister about the Industrial Revolution, signifying so much growth and progress all within just 150 years, and all because of him. Picking up a handful of the accumulating black carbon, she retorts that this is all because of him too. Concerned, he reminds her that she said it was no big deal when the first humans were releasing carbon, to which she counters that that was such a small amount. Behind their conversation, the industrial music has transformed itself into the stirrings of a storm. Foss's dancers are caught in the swirl of wind and slowly begin circling the fabric of community, their flags now suggesting increasingly high winds, thunder, and lightning. Foss asks his sister what's going on. She says she's seen this happen on this planet before; the climate is changing. He asks why, to which she retorts that it is the carbon cycle; he has disrupted the natural carbon cycle. The storm continues to build in intensity and speed until, at its climax, several of the circling flag bearers are knocked off course by the storm and tear through the fabric of the community, ripping the strands of paper in a cacophony of destruction. One of the weavers is injured and falls center stage. Foss lurches to catch her and holds her head in his arms. The music stops. The dust settles. Everyone gathers around, watching in silence. Foss looks up to his sister and asks, "What now"? That pose is held while one of the performers steps forward to address the audience, "This is where we are now as a human community. Our use of fossil fuel energy is impacting our climate and those who did the least to cause it are being hurt by it the most. In the face of these challenges, how do we want to prepare? What story do we want to tell for our city? How do we plan to get from this point in history to a resilient future? That part of the story will now be told".

What follows, in Act Two, is a number of youth-authored solutions to youth-identified social stresses and climate threats at the local level (Fig. 1.6). These are enacted as short skits, one right after the other. This can take several different forms, as will be described in more detail in Chap. 3. In the transitions between the skits, the entire cast chants, "Bounce forward, rebound, that's my resilient town." The final song and dance celebrates what was accomplished and strengthens the community's resolve to put these solutions into action.

Fig. 1.6 Youth performers enact skit for Act Two of *Shine*. Photo by © 2016 Steven Sutton, DUOMO

Notes

1. Rockefeller Foundation, "100 Resilient Cities," September 1, 2016, http://www.100resilientcities.org/#/-_/.
2. Rockefeller Foundation, "About Us| 100 Resilient Cities," September 1, 2016, http://www.100resilientcities.org/about-us#/-_/.
3. Michelle Fox, Marcus Moench, and Rachel Norton, *Beyond Resilience* (Institute for Social and Environmental Transition-International, 2015).
4. Bruce Goldstein et al., "Narrating Resilience: Transforming Urban Systems Through Collaborative Storytelling," *Urban Studies* 52, no. 7 (May 2015): 1300.
5. Ezra Markowitz, Caroline Hodge, and Gabriel Harp, "Connecting on Climate: A Guide to Effective Climate Change Communication" (New York: Center for Research on Environmental Decisions, Columbia University, 2014), 24.
6. D. Abrahamson, "Embodied Spatial Articulation: A Gesture Perspective on Student Negotiation between Kinesthetic Schemas and Epistemic Forms in Learning Mathematics.," in *Proceedings of the 26th Annual*

Meeting of the North American Chapter of the International Group for the Psychology of Mathematics Education, vol. 2 (Toronto, Ontario: Preney, 2004), 791–97; C. Fadjo, M. Lu, and J.B. Black, "Instructional Embodiment and Video Game Programming in an after School Program" (World Conference on Educational Multimedia, Hypermedia and Telecommunications, Chesapeake, VA, 2009).

7. Jan Cohen-Cruz, *Local Acts: Community-Based Performance in the United States* (New Brunswick, NJ: Rutgers University Press, 2005), 99.

8. Susan Kattwinkel, *Audience Participation: Essays on Inclusion in Performance* (Westport, Conn: Praeger, 2003), 186.

9. Elke Weber, Daniel Ames, and Ann-Renee Blais, "'How Do I Choose Thee? Let Me Count the Ways': A Textual Analysis of Similarities and Differences in Modes of Decision-Making in China and the United States," *Management and Organization Review* 1, no. 1 (2005): 87–118.

10. Hannah Reid et al., *Community-Based Adaptation to Climate Change*, Participatory Learning and Action 60 (London: International Institute for Environment and Development, 2009).

11. Wendy Arons and Theresa May, eds., *Readings in Performance and Ecology* (New York, NY: Palgrave Macmillan, 2012), 4.

12. Cheryll Glotfelty and Harold Fromm, eds., *The Ecocriticism Reader: Landmarks in Literary Ecology* (Athens, Georgia: University of Georgia Press, 1996).

13. Greg Garrard, *Ecocriticims*, 2nd ed. (New York: Routledge, 2011), 16.

14. James Thompson, *Applied Theatre : Bewilderment and beyond* (Oxford UK: Peter Lang, 2008); Monica Prendergast and Juliana Saxton, *Applied Theatre: International Case Studies and Challenges for Practice* (Bristol; Chicago: Intellect Ltd, 2010); Michael Rohd, *Theatre for Community Conflict and Dialogue: The Hope Is Vital Training Manual* (Portsmouth, NH: Heinemann, 1998); Philip Taylor, *Applied Theatre: Creating Transformative Encounters in the Community* (Portsmouth, NH: Heinemann Drama, 2003).

15. David Pammenter, "Theatre as Education and a Resource of Hope: Reflections on the Devising of Participatory Theatre," in *Learning Through Theatre: The Changing Face of Theatre in Education*, 3rd ed. (New York: Routledge, 2013), 83.

16. Richard Schechner, "Foreword: Fundamentals of Performance Studies," in *Teaching Performance Studies*, ed. Nathan Stucky and Cynthia Wimmer (Carbondale, IL: Southern Illinois University Press, 2002), ix.

17. Ibid., xi.

18. Shannon Jackson, *Social Works: Performing Art, Supporting Publics*, 1st ed. (New York: Routledge, 2011), 211.

19. Ibid.

20. Tony Fry, *City Futures in the Age of a Changing Climate* (New York City: Routledge, 2015), viii.
21. Louise Chawla, *Growing Up in an Urbanizing World* (New York: Routledge, 2001).
22. Maria Ojala, "Regulating Worry, Promoting Hope: How Do Children, Adolescents, and Young Adults Cope with Climate Change?," *International Journal of Environmental & Science Education* 7, no. 4 (2012): 537–61.

References

Abrahamson, D. "Embodied Spatial Articulation: A Gesture Perspective on Student Negotiation between Kinesthetic Schemas and Epistemic Forms in Learning Mathematics." In *Proceedings of the 26th Annual Meeting of the North American Chapter of the International Group for the Psychology of Mathematics Education*, 2:791–97. Toronto, Ontario: Preney, 2004.

Arons, Wendy, and Theresa May, eds. *Readings in Performance and Ecology*. New York, NY: Palgrave Macmillan, 2012.

Chawla, Louise. *Growing Up in an Urbanizing World*. New York: Routledge, 2001.

Cohen-Cruz, Jan. *Local Acts: Community-Based Performance in the United States*. New Brunswick, NJ: Rutgers University Press, 2005.

Fadjo, C., M. Lu, and J.B. Black. "Instructional Embodiment and Video Game Programming in an after School Program." Chesapeake, VA, 2009.

Fox, Michelle, Marcus Moench, and Rachel Norton. *Beyond Resilience*. Institute for Social and Environmental Transition-International, 2015.

Fry, Tony. *City Futures in the Age of a Changing Climate*. New York City: Routledge, 2015.

Garrard, Greg. *Ecocriticims*. 2nd ed. New York: Routledge, 2011.

Glotfelty, Cheryll, and Harold Fromm, eds. *The Ecocriticism Reader: Landmarks in Literary Ecology*. Athens, Georgia: University of Georgia Press, 1996.

Goldstein, Bruce, Anne Wessells, Raul Lejano, and William Hale Butler. "Narrating Resilience: Transforming Urban Systems Through Collaborative Storytelling." *Urban Studies* 52, no. 7 (May 2015): 1285–1303.

Jackson, Shannon. *Social Works: Performing Art, Supporting Publics*. 1st ed. New York: Routledge, 2011.

Kattwinkel, Susan. *Audience Participation: Essays on Inclusion in Performance*. Westport, Conn: Praeger, 2003.

Markowitz, Ezra, Caroline Hodge, and Gabriel Harp. "Connecting on Climate: A Guide to Effective Climate Change Communication." New York: Center for Research on Environmental Decisions, Columbia University, 2014.

Ojala, Maria. "Regulating Worry, Promoting Hope: How Do Children, Adolescents, and Young Adults Cope with Climate Change?" *International Journal of Environmental & Science Education* 7, no. 4 (2012): 537–61.

Pammenter, David. "Theatre as Education and a Resource of Hope: Reflections on the Devising of Participatory Theatre." In *Learning Through Theatre: The Changing Face of Theatre in Education*, 3rd ed., 83–102. New York: Routledge, 2013.

Prendergast, Monica, and Juliana Saxton. *Applied Theatre: International Case Studies and Challenges for Practice*. Bristol; Chicago: Intellect Ltd, 2010.

Reid, Hannah, Mozaharul Alam, Rachel Berger, and Terry Cannon. *Community-Based Adaptation to Climate Change*. Participatory Learning and Action 60. London: International Institute for Environment and Development, 2009.

Rockefeller Foundation. "100 Resilient Cities," September 1, 2016. http://www.100resilientcities.org/#/-_/.

———. "About Us | 100 Resilient Cities," September 1, 2016. http://www.100resilientcities.org/about-us#/-_/.

Rohd, Michael. *Theatre for Community Conflict and Dialogue: The Hope Is Vital Training Manual*. Portsmouth, NH: Heinemann, 1998.

Schechner, Richard. "Foreword: Fundamentals of Performance Studies." In *Teaching Performance Studies*, edited by Nathan Stucky and Cynthia Wimmer. Carbondale, IL: Southern Illinois University Press, 2002.

Taylor, Philip. *Applied Theatre: Creating Transformative Encounters in the Community*. Portsmouth, NH: Heinemann Drama, 2003.

Thompson, James. *Applied Theatre : Bewilderment and beyond*. Oxford UK: Peter Lang, 2008.

Weber, Elke, Daniel Ames, and Ann-Renee Blais. "'How Do I Choose Thee? Let Me Count the Ways': A Textual Analysis of Similarities and Differences in Modes of Decision-Making in China and the United States." *Management and Organization Review* 1, no. 1 (2005): 87–118. doi:1740-8776.

CHAPTER 2

Design of *Shine* as a Method for Engagement

Abstract This chapter includes performance materials for various levels of youth-engagement for use by teachers, community organizers, or faith leaders. The "Notes from the Composer," and the "Notes from the Choreographer" both have corresponding videos (links provided), of which these are a slightly adjusted written transcription. These video recordings are educational tools to prepare either the facilitator, student, or scholar to more deeply appreciate the music and the movement within this performance. Materials for building curriculum included in this chapter can be adapted or used as they are or be scaled to match the age of the youth participating. The sample exercises and games can be used to prepare youth for creative expression, to build community, to expand their expressive range, and for fun.

Keywords 100 Resilient Cities · Youth engagement · Applied theatre
Climate change · Energy · Resilience · Script · Music · Composer
Choreography · Choreographer · Climate scientist · Curriculum
Exercises and games

This chapter includes materials for the use of *Shine* for various levels of youth-engagement as facilitated by teachers, community organizers, or faith leaders. These materials include the script, notes from the composer, choreographer, and a partnering climate scientist; materials for building a curriculum; and sample exercises and games. Combined, these elements constitute the design for this project being studied through this book. In the Appendix is a "Choreography Worksheet with Counts and Lyrics for Each Dance in *Shine* with Open Spaces for Choreographer Notes" that will serve as a useful tool for anyone facilitating the movement for *Shine*. The Notes from the Composer, and the Notes from the Choreographers both have corresponding videos (links provided within the text), of which these are a slightly adjusted written transcription. These video recordings are educational tools to prepare the facilitator, student, or scholar to appreciate the music and the movement within this performance more deeply. The materials for building curriculum included in this chapter can be adapted, used as is, or can be scaled to match the age of the youth participating in this project. The sample exercises and games can be used to prepare youth for creative expression, to build community, to expand their expressive range, and for fun.

- Script (including music files for songs—both with and without vocal tracks)
- Notes from the Composer
- Notes from the Choreographers
- Interview with Partnering Energy Scientist
- Materials for Creating Curriculum—adaptable by K-12 schools, youth, and faith organizations
- Sample Exercises and Games

I wrote this script in consultation with various specialists. The process was approached holistically: the story, movement, and science came together through active collaboration. For instance, when *Shine*'s choreographer Arthur Fredric was in Boulder for an artistic residency at CU, we met in a dance studio with Evolutionary Biologist Nicole Berger to create the section on photosynthesis. Our goal was to condense the complicated process of this phenomenon clearly and accurately in a way that could also serve as the basis for movement by the youth portraying ancient plants. As Berger described the various parts of photosynthesis, we identified simple names for each part, and then explored movement that could portray each part, such as plants gathering water from their roots. We worked together until: I could write it, Fredric felt he could effectively choreograph photosynthesis, and Berger felt it was being accurately portrayed. Once the script was completed I shared it with

Joshua Sperling to ensure that not only the science of climate and energy was being accurately represented but also that the concept of resilience as applied to city planning was being accurately represented. His only comment was to change the transitional chant in Act Two from "Bounce back, rebound, that's my resilient town," to "Bounce forward, rebound, that's my resilient town," to accurately capture the underlying goal of resilience planning at the city level. His change represented the desire to advance after mistakes are made or disasters occur, such that city planners learn from these setbacks and incorporate the lessons learned into improved designs that attempt to alleviate the vulnerability that initially caused the setback (see http://www.insidethegreenhouse.org/shine).

Script with Stage Directions and Sound Cues

All performers besides Sol should be wearing black pants and shirts with no logos (can turn shirts inside out, be sure to cut off tags). Sol and Foss begin wearing their costumes. The Seed Sower is dressed in black and is wearing her/his sash that holds tissue paper seeds. (For the full performance see: http://www.insidethegreenhouse.org/shine)

Preset:
Brown cloth along back of the stage
Dinosaur costume or sculpture (if using any, can be physically represented by performers)
Sticks for humans
Red cloth for fire
Black cloth bag with black tissue-paper carbon in it
Sash with seeds for Seed Sower
Waist sash/skirts and head wraps for Harvesters
Newspaper plants for Harvesters
Sunglasses and sashes for Foss Folks
Paper weaving rolls of community for Harvesters
Poles with Fossil Fuel flags
Black carbon for Foss (in shoulder strap bag or his pockets)
Big Balloons for ending (optional)

Act 1:
SOUND CUE: *Long Time Comin'* (https://soundcloud.com/shinemusical/long-time-comin)

(Primordial sounds, then 2 counts of 8 introductory instrumental music for entrance of ancient plants and animals)

Long time comin', a long time a comin' comin', long time comin' along
Long time comin', a long time a comin' comin', long time comin' along
Long time comin', a long time a comin' comin', long time comin' along
Long time comin', a long time a comin' comin', long time comin' along
(7 counts of 8 rhythmic music continues low underneath to maintain the mood, then a bit more of just primordial sounds, fades out—Sol should enter and speak once the sung portion of the song is complete)

Sol: The weirdest thing happened about 300 million years ago. I was just shining down on this planet like I do on all my planets—I have eight—and then these ancient plants and animals start doing this musical number. Yes, a musical number. Weird, right? I think, yeah, I'm down with this. Life is getting more and more animated on this planet. I can adjust. (Sol watches as the ancient plants reach and contract, moving around. Ancient animals are along the sides moving as their animal would move, looking at the plants hungrily.) This dance number is kind of interpretive, but it looks like the ancient plants are taking in CO_2 from the atmosphere to photosynthesize my energy to store it up as carbon.
I think they are dancing the different parts of photosynthesis.
(Ancient plants reach towards the Sun for her energy.)
They need my energy from the Sun.
(Ancient plants enact taking up water from their roots.)
They take up water through their roots.
(Ancient plants enact having excited cells.)
That excites their cells.
(Ancient plants enact breathing.)
They breathe in CO_2 from the atmosphere.
(Ancient plants grow.)
They grow.
(Ancient animals move in and begin feasting on the ancient plants.)
They get eaten by ancient animals; so now the animals have my energy in them too.
(All die slowly and dramatically, landing together in a clump so the brown cloth can completely cover everyone.)
And then they all die, both the plants and the animals.
And this keeps happening over and over again.

(4 stage hands, each at a corner of the brown cloth, slowly pull brown cloth over all of their dead bodies. Foss should sneak under the brown cloth unseen by the audience.)

They get covered up by hundreds, sometimes thousands, of feet of mud and rock and sand. They turn into fossils. This goes on for millions of years. (Enter Foss from beneath the brown cloth, unseen by Sol, and eventually bumps into her) I assume they're just down there decomposing. And then, Wham! This person bumps into me. And I say, Whoa. Who are you?

(Once under the cloth, the performers representing ancient plants and animals should take off their animal or plant costumes, and deposit them in the center under the brown cloth and wait to exit from beneath cloth. Stage hands hold the cloth down so they are not revealed to the audience.)

Foss: I'm your brother.

Sol: I don't have a brother.

Foss: You do now.

Sol: What? How can that be?

Foss: Yeah, I'm another form of energy. Like in your family, your brother.

Sol: What do you mean? I'm energy. I'm the sun. For your information, energy can neither be created nor destroyed. So, I can't have a new energy brother.

Foss: Back to school, sis. Energy can't be destroyed, but it can be transformed from one form into another. All those fossils of decomposing plants and animals that stored all that solar energy from you, remember them? They formed fossil fuels. Pow! I was born. I'm a new form of energy on this planet. You could call me fossil fuels, but I'd really prefer something a bit flashier, like Foss! What do you think?

Sol: What have I done?

Foss: It's no big deal, it's not like I'm going anywhere. I'm mostly stuck under the ground. See? All those bumps underground? That's most of me.

(Performers beneath the brown cloth can now begin to roll out from under the cloth to the sides of the stage, try not to let the audience see the animal and plant costumes that are remaining under the brown cloth. Once all the performers exit from beneath the cloth, stage hands bunch it all together and remove from stage area, being careful not to expose any of the animal or plant costumes.)

Sol: Okay, this really is not what I expected, but I could get used to having some family. I guess the company's not bad. So, new energy brother, come with me. (Sol begins making repeated gestures of shining.)

Foss: What do we do now?

Sol: I just shine. Millions of years go by, and I just keep on shining. That's what I do.

Foss: What do I do?

Sol: I don't know. What can you do?

Foss: That's just it, I don't know, but I was hoping for a bit more action. (Sighs, then asks restlessly,) What time is it?

Sol: It's Triassic Period, about 231 million years ago.

(Have ensemble represent a dinosaur crossing the stage.)

Foss: Whoa. What is that?

Sol: Those are new. Dinosaurs, let's call those dinosaurs.

Foss: Cool. (The dinosaur exits into extinction. Foss pauses, slumps.) What time is it now?

Sol: About 2.8 million years ago.

(Enter two humans, each holding a stick, another two people behind them—one holding a flame and the other ready to toss black tissue paper as carbon once fire is lit.)

Foss: What's that?

Sol: That's a new one too. Humans. Yep. Let's call them humans.

Foss: What is that human doing?

Sol: I don't know. Wait. That human is breaking up sticks and rubbing them together.

Foss: What's that red stuff?

Sol: Smart. Those humans are using energy from me that's stored up in wood from a tree to make heat energy. Clever. These humans are fun to watch.

(Sprinkle carbon over the fire.)

Foss: What's that black stuff?

Sol: Oh, that's just carbon being released into the atmosphere. But there's not very much, so it's not a big deal.

SOUND CUE: *Harvest* Song/Foss Folks (https://soundcloud.com/shinemusical/harvest)

(The following conversation happens quickly over the music so it concludes before the singing begins.)

Foss: What's that noise?

Sol: Oh, I forgot to tell you, you're going to like this if you want some action. Sometimes life on this planet does musical numbers. The ancient plants and animals were just doing one about 297 million years ago. It's cool. Wanna join in?

Foss: Not really my style. I'll wait for the next one.

Sol: Could be a while.

(2 counts of 8 introductory instrumental music for entrance of Harvesters holding newspaper plants.)

We plant together standing side by side
We reap together with our arms open wide
We work together with the seeds that we sow
We feed each other with the foods that we grow

We plant together standing side by side
We reap together with our arms open wide
We work together with the seeds that we sow
We feed each other with the foods that we grow

(4 counts of 8 introductory instrumental music for entrance of Foss and the Foss Folks, who knock the newspaper plants over in the hands of the Harvesters)

We don't tend sheep anymore
We don't harvest wheat anymore
Sister don't be such a bore
We get our food from a store
We don't sleep at night anymore
Cause it ain't such a fright any more.
Sister don't be such a bore
Get out and dance on the floor

You must agree that your ways are wasteful
You must agree that my path is more tasteful

I bought the shoes on my feet
Drive my car down the street
It's hard to believe that we're related
Your ways are antiquated

(Seed Sower can act out or sing these lines)
It's the harvest party, let's not fight
Or waste our time on whose wrong or right
Brother, sister now let's get along
Let's weave together the words of these songs

(Mash up of their two songs, Part A and Part B below.)

Part A
We plant together standing side by side
We reap together with our arms open wide
We work together with the seeds that we sow
We feed each other with the foods that we grow

We plant together standing side by side
We reap together with our arms open wide
We work together with the seeds that we sow
We feed each other with the foods that we grow

Part B

We don't tend sheep anymore
We don't harvest wheat anymore
Sister don't be such a bore

We get our food from a store
We don't sleep at night anymore
Cause it ain't such a fright any more
Sister don't be such a bore
Get out and dance on the floor

We don't tend sheep anymore
We don't harvest wheat anymore
Sister don't be such a bore
We get our food from a store
We don't sleep at night anymore
Cause it ain't such a fright any more
Sister don't be such a bore
Get out and dance on the floor
Sol: Are you always going to spoil everything?

Foss: Come on, you had fun. Admit it, you liked my funkier beat.

Sol: It was alright. Okay, I guess it was kind of fun.

Foss: You guess? You loved it. That number was dragging before I came in. Come here, sis. Bring it in. (They embrace and laugh.) Hey, we're bonding.

(Humans enter, some arm in arm, all getting to know each other.)

Sol: I guess we are.

Foss: It's nice.

Sol: Yeah, it's nice. (Sit together.) Look, the humans all seem to be settling into a clump.

Foss: I think they're bonding too.

Sol: You're right; they're forming a community.

Foss: (Reaches for one of the paper stalks left by the Harvesters and regards it.) They can do that now since they figured out how to make enough food to stay in one place. I wonder what they call themselves.

Humans: We hereby declare ourselves the city of (name of your city)!

(Can create some comic bit here that expresses what is unique or iconic about your city.)

(Harvesters collect the paper weaving rolls of community for the Weaving song, can do stylized movement here to show them becoming a human-powered machine, a human loom. Performers may need to get people from the audience to be part of the loom if you don't have enough performers. Put audience members in the more passive, receiving sides of the loom.)

Sol: Hey, it looks like they're working together.

Foss: It drives me crazy how everything on this planet happens so slow. Whatever they are trying to do is going to take forever just using human energy. There's probably a faster way to do this.

Sol: This is good. They're figuring it out. Look, they're working together making a machine to weave fabric, a loom. They're going to weave who they are as individual humans into a community using a human loom.

Foss: This might be where I could come in. My purpose. I can help these machines and (name of your city) go faster, with more power!

Sol: Careful. You're an energy form. They're just humans. They might not be able to handle you. You've got millions of years' worth of my solar energy packed into you.

Foss: Relax. I'm just trying to help them.

Sol: But you don't know what's going to happen if you let loose.

Foss: Progress, that's what going to happen. And progress is not such a bad thing. These humans seem to want it. I'm right beneath their feet—coal, oil, and natural gas. Look at all the toil and struggle these humans have to go through just to meet their basic needs to eat and be warm. You've seen it; they have so much potential. They're clever. Just imagine what they could create with my power to fuel their ideas.

Sol: Slow down. These folks seem to have found a really nice balance just using solar energy and biomass.

Foss: Yeah, but that's not for everybody. Let's see what they want.

Sol: Somebody could get hurt.

Foss: And somebody could be jealous of her brother.

(Both siblings turn from each other in a huff. Foss goes to his Foss Folks. Sol watches over the Harvesters as they weave, shining from the side. Once the singing starts, weaving should begin slowly.)

SOUND CUE: *Weaving* (https://soundcloud.com/shinemusical/weaving)
 Over thread and under strand
 Over time we understand
 Fibers will combine to be
 The fabric of community

 Ancestry and history
 Cloth to warm us, hold and form us

 Sun is constant always there
 Rays of light weave through the air
 Come on out and sow the seeds
 Simply we can meet our needs

 Ancestry and history
 Cloth to warm us, hold and form us

 Over thread and under strand
 Over time we understand
 Fibers will combine to be
 The fabric of community

(At some point when the weaving is complete, everyone tilts the entire fabric to reveal it to the audience. Hold this in place during the *Progress*/Storm.)

SOUND CUE: *Progress*/Storm (https://soundcloud.com/shinemusical/progress)
(1 count of 8 introductory instrumental music for entrance of Foss Folks carrying fossil fuel flags. Foss releases fistfuls of black tissue paper to represent the release of carbon throughout this song.)

 Come with me to fuel the world, I'm looking for coal
 Come with me to fuel the world, I'm looking for oil
 Come with me to fuel the world, I'm looking for gas
 Fuel to meet increasing needs to move fast
 Just you and me, I'm energy

(Some of Foss's people use their flag poles like shovels and begin digging into the ground to unearth the fuels. Others use their flag poles to mime thrusting their flag into the ground to claim this land for progress. The driving rhythm of the machines begin to slow. Foss and Sol have the following conversation immediately over the music. They need to speak somewhat quickly to match the coming of the storm caused by the excessive release of carbon from the use of fossil fuels.)

Foss: Sis, look! It's like an Industrial Revolution! So much growth and change in just 150 years. All because of me!

Sol: (Sol picks up some carbon) And this is all because of you too. Look at all this carbon you've released.

Foss: Yeah, but you said it was no big deal when the humans were burning wood.

Sol: That was such a small amount; look at all this.

Foss: (Listening to the soundtrack of the storm.) Hey sis, what's that noise?

LIGHTING CUE: (Could do a flick of lights to make it look like lighting, could get darker, should build with the song, continue with lightning-like effects.)

Sol: I've seen this before. The climate is changing on this planet again.

Foss: What? Why?

Sol: The carbon cycle, you disrupted the natural carbon cycle.

(Foss and Foss Folks get caught up in the winds of the storm, wave their flags out of control, lose balance, and eventually crash into the fabric of community, tearing it and destroying it completely. One of the Harvester/Weavers portrays being hurt by this, swoons, and falls in the center of the stage. Foss sees this and catches her. He holds her in his arms and looks up to the others who have gathered in a half circle behind him. He takes in the consequences of his actions.)

Foss: What now?

(All freeze in this tableau. Two performers step forward and deliver the following lines directly to the audience.)

Performer 1: This is where we are now as a human community. Our use of fossil fuel energy is impacting our climate and those who did the least to cause it are being hurt by it the most.

Performer 2: In the face of these challenges, how do we want to prepare? What story do we want to tell for our city? How do we plan to get from this point in history to a resilient future? That part of the story will now be told.

Act 2:

(All performers break from the tableau, get into their skit groups, and begin chanting "Bounce forward, rebound, that's our resilient town" while clapping.)

(Present youth-authored skits that dramatize their solutions to challenges to resilience that they have identified in their own community. Skits can be about 1–2 minutes in length and should actively communicate these solutions in a creative and embodied manner.)

Sample Skit by youth participants in Boulder, Colorado, USA, for the performance at the National Center for Atmospheric Research, described in greater detail in Chap. 3.

(this was created and performed by the two lead characters who played the roles of Sol and Foss along with three other cast members from the ensemble, Sol and Foss retained their roles for this skit)

Sol: Well, I think we should replace all energy created by using you (referring to Foss), and replace it with me.

(audience applauds)

Foss: (directly to the audience) Really? I know what I did back there wasn't ideal, but you can't just get rid of me. I've done so much for these people. I've given them jobs, a home, electricity. You can't just get rid of me. For example, (enter Billionaire and Poor Old Woman), you say you want to replace everything with renewable energy. Well what if you replace everything with solar panels? Mr. Billionaire over here is like "oh yeah, I can afford that. I have money," but this poor old lady who can barely afford food, how do you think she is going to be able to afford solar panels? Those are expensive.

Sol: Okay, so we won't start with homes. (Ensemble members act out each of the following solutions.) We can start in the streets, like street lights—they can be charged with solar energy by day, and then at night they'll be powered by solar energy to light the streets. We can also have building regulations that are Green Star and eco-friendly.

Foss: Good idea. Anything else?

Sol: Hmm. Oh, I know. We can teach elementary school students about solar and renewable energy, and it can be youth-led.
(ensemble members enact two students listening to a student teacher)

Student Teacher: And that's what you kids need to learn about solar energy.

Students: Go solar!

Foss: This is really coming together. We finally made a good community.

Sol: I think we have. (They jump up and do a high five.)

(Additional examples of skits that vary in style and complexity can be found through the descriptions of the tour in Chap. 3, along with descriptions of how they were created.)

(To transition between each skit, chant, "Bounce forward, rebound, that's our resilient town" either once or twice. Optional: Could have each group draw some representation of their solution on a large balloon that they integrate into their skit, or that someone else holds above or near their skit while they perform it. After the final skit is over, have all performers get into place for the final dance number, *Shine*, while chanting:)

> Bounce forward, rebound
> That's our resilient town
> Bounce forward, rebound
> That's our RESILIENT TOWN!
> *If each group has a balloon, throw it back to the stagehands while they dance to the following song.

SOUND CUE: *Shine* (https://soundcloud.com/shinemusical/shine)

> (4 counts of 8 introductory instrumental music)
> Turn around touch the ground

Til a new idea is found
Look up, look down, shake up your town
Swish your feet, repeat
Right down our main street
Light bright feels right
Run for fun in the sun and
Shine shine shine shine
Shine shine shine
Shine shine shine

Turn around touch the ground
Til a new idea is found
Look up, look down, shake up your town
Swish your feet, repeat
Right down our main street
Light bright feels right
Run for fun in the sun and
Shine shine shine shine
Shine shine shine
Shine shine shine

(If using balloons, release big balloons for the performers and audience to bounce among each other to demonstrate their active support of youth-authored ideas. Best to follow the applause immediately with another song chosen by the performers to keep the energy up as they play with the balloons and mingle with the audience.)

THE END

Notes from the Composer

Tom Wasinger, three-time Grammy winner, on the composition of the songs (Video of interview Inside the Studio and recording of each song referred to below is available at: http://www.insidethe-greenhouse.org/shine/shine_music.html)

Long Time Comin'

I was asked to make a song about life coming out of the primordial soup during the beginnings of life on Earth. I started this piece out with an East African drum—very different from the West African *djembe* drum.

This East African drum has many pitches and invokes both the spirit of the tree used to make the body of the drum, and the spirit of the animal used to make the head of the drum. A dried piece of the heart of that same animal is put inside so the spirit of the animal is in the drum. This gives an entirely unique sound, primordial and mysterious. I layered samples of a little wooden instrument that I just call a frog—a small percussion instrument made to evoke the sounds of insects and amphibians in the wild. Next came samples of jungle birds. I combined all of these for the beginning of this piece to create this vibrant teaming of life. For my vocals, I used the lower parts of my range. Deeper yet, I actually "played" the closet in my studio. Yes, I have strung bass guitar strings onto my wooden closet. Because the closet is so large, it can rumble the whole room, and thus support the fundamental of that note. And voila, there we had it: life springing out of the darkness of our history.

Weaving

When creating the songs, I used sounds that are connected to the theme of the piece. I was travelling in Turkey years ago and visited a community building especially dedicated to women's weaving. While many women wove, the looms clanked and swooshed. Taking bits and pieces of this recorded sound, I created a loop of the sounds. One of the snap sounds had an inherent pitch, which gave me a starting point in terms of tonality and a melody that I could add to. However, it would be too repetitious to have just that. So, I found another loop that matched the first, which was challenging in terms of matching the tempo and the tonality. Those two combined are the bed for the first verse. By the second verse we add vocal harmony to build up the song in richness. Another sound I introduce in this piece is a guitar swell that I recorded and turned around backwards—a common recording technique that even the Beatles used. By using those various elements—joined with the refrain and the verses—I had the piece built and become a deeper and deeper journey into the world of looms all working together to weave fabric.

Progress/Storm

This is a piece that was composed to exhibit industry on parade— man's use of energy to build infrastructure and transportation that led to the Industrial Revolution. It had to convey a feeling of pushing forward, of driving into the future. Thus, I wanted to start with something that sounds very mechanical. I used a loop of a recording of a steam engine. To that loop, I added the sound of a nail gun. To that I added a sample of a basketball hitting a hollow wooden floor. Next the sound you get when you open a jar of peanuts, and the air escapes. Putting all those together created the bed of the piece. For a melodic element, I brought in an instrument called an *inombiganda*, which looks like a cello but is more related to a guitar, as it has frets. It is from a family of stringed instruments from Western Europe, popular before and during the Baroque Era. I used it to create single notes that were deeply resonant and the tiniest bit menacing. By having four tracks of that instrument going at the same time, it sounds like an ensemble. This serves as the foundation for the vocal, "I'm looking for coal, oil, and gas," in the piece. The over-arching drive for this song is this human urge to aggressively delve into the depths of the earth to fuel a lifestyle that led to the Industrial Revolution. Oh, and I forgot the munchkin voices. I recorded my voice saying, "big big rig, big rig swig" over and over again, and "munchkinized" it to create this background chant. Listen for it.

Through a musical cue in this song, Foss becomes aware of the fact that perhaps all is not well with this abundant use of energy. We hear the steam engine grinding to a halt and the other tracks likewise starting to slow down and almost come to a stop. In an ominous transition, the sounds slowly regain speed again but backwards. Claps of thunder indicate a storm is swelling. Here, I added the sound of a wind harp. I play this "instrument" by putting strings on a tree outside my mountain house to record the sound of the wind with an audio pick up attached to the tree. The sound keeps getting higher and higher as the wind blows through the strings. Towards the end, I layer a recording of water running through a pipeline that runs very close to my house. On top of that, we hear cowbells clanging—actual cowbells from real cows in Switzerland. Added to this cacophony near the end of the piece is music of the Marksiphone, which is in the zither family. It is spring loaded, played by little hammers, and sounds like a boneshaker. This eventually builds up into the crescendo of destruction. Foss's miners are tossed and turned by the raging storm and rip through the woven fabric of community. The final crashing sound we hear as an injured weaver falls into Foss's arms, is a recording of a giant metal door slamming shut adjacent to a large stone room in a castle in Britain.

Shine

This final song is intended to send you home feeling good and humming this tune; we wanted it to be a big celebration and a great dance number. It is a take-off of a typical funk rhythm. To be honest, we were very inspired by Bruno Mars. This song has a typical bass line that is sung with vocals. It has guitar add ins, and a clavinet keyboard instrument that is in 1970s Funk music. All the melodic parts fit together by themselves. I added together many sounds to fill out the rhythm section, including four different samples for the snare sound alone. Same thing with the base drum sounds on one and three. I was using the chair I sit on in my studio, which is a *kahon* (a hollow wooden box), a kick drum from a drum kit, and a sample of a gigantic Indian drum as big as a table. All of those together gave us our kick sound.

NOTES FROM THE CHOREOGRAPHER

Arthur Fredric, Master Teacher with the New York City National Dance Institute and former Broadway performer, and Lisa Denning, Assistant Choreographer, former professional actor and dancer (Video of interview Inside the Studio available at: http://www.insidethegreenhouse.org/shine/shine_choreography.html)

Fredric: I'm Arthur Fredric the director and choreographer of *Shine* for the version that we did in Connecticut and New York. I've been a teaching artist with the National Dance Institute (NDI) for twenty years. NDI is the brain child of the great Jacques d'Amboise who was one of the top classical dancers for the City Ballet for twenty-seven years. His thought was, why just me? Why just these elite ballet dancers experiencing the wonder of dance and movement? So, he literally just went into public schools and started teaching. NDI is the gold standard in arts education in the USA and around the world. Everybody looks to NDI, their teaching methodologies, their enthusiasm, their high intelligence about how to approach teaching, and how to work with ordinary people and getting them excited and moving, and invested in the beauty and the magic of the arts.

Denton: One of the most important things in working with children is giving them a tremendous amount of respect and also presenting things in a very clear simple way. Movement should be presented in achievable chunks so that they can feel good about themselves and not get overwhelmed or frustrated, but rather so they can gain a great sense of achievement. A sense of humor is always important so movement feels like play to kids. It should be hard work, but it needs to be joyful and hard. You need to raise the bar but in a way that makes them feel really good about themselves.

Fredric: When you're working with people that have never been on stage before, you're teaching them how to just rhythmically, artistically, and dynamically get into their body and create simple, everyday things that we can all do. Once we learn how to connect to rhythm and music—things that all of us have at our disposal—we begin to express in the universal language. When you clap beats, you can feel the room working together. So, I "tune the room," by getting them clapping together, tapping their toes all together, even just lifting their knees at a certain level, or changing feet together. And once you get everybody "tuned up" and realizing they know their right foot from their left, they know how to hit a down beat, and it feels good in their bodies, then they're in a position to learn. That takes a certain amount of repetition, but you don't want repetition that deadens awareness. You want repetition during which you're giving notes and you're encouraging them. The movement is getting better and better until there's a group fire; until you're building a bonfire of achievement.

In every class, there's going to be different levels of achievement. To get people comfortable, establish that everybody is cared for. Then when you see one person who's getting everything right, start by singling that person out and having that person model excellence. When everyone sees that excellence and cheers, they understand "Whoa, that's really good." Then other people will aspire to that excellence. Then you can start modelling excellence with a person who's struggling. Say this person is starting each step with his left foot when he should to be starting with his right. Single out that person and say, "Let's just get the first count. Here's your right foot; take the step on one. Five, six, seven, eight, and right foot." And then he does it right, and "Yes! You did it!"

Now you're correcting while taking all that care, and everybody can see that the person is achieving. So, you're not just saying "Oh good" when it's not good. You're actually helping the person achieve it, and then they can start feeling that they're achieving excellence also.

Denton: Then it is important to add the intention or the meaning to the movement. What is inspiring you to move? You're telling a part of a story, and your character is really important. If you're doing the Harvest song in *Shine*, you want to feel Mother Earth. You want to feel like you're connected to the Earth and how important that is. If you're dancing as fossil fuel, you want to feel energy and how powerful it is. Inspiring young people to "become" the movement and understand the real meaning from within gives them a feeling of "Yeah I can do that." Fossil fuel is intense, "Come on, go!" You can really feel it, and then the kids can really get into it. Then you are inspiring them through what the character is or what the intention of the movement is in that moment.

Fredric: My advice for someone approaching *Shine* would be first to understand this is a piece that takes such a serious foreboding topic, and it adds humor, dialogue, pathos, possibilities, and hope. So much of that is invested in this piece. Climate change is such a serious topic today, but really what this piece is about is youth-led engagement. Youth with the proper leadership can shine. Actually experiencing it and getting up and physically participating is wonderful; it's almost a salve. Any of our concerns, doubts, or worries just seem to disappear in the work, in the accomplishment. Getting up and participating and having scientists participate with the actual people who are sharing their concerns can have a wonderful effect on the entire room and the community.

When kids experience joy through their bodies, there's actually a great deal of self-respect that they gain. They can feel joy. They can feel important seeing they should take up some space in the world. All the adults who spent that time with the magic of a child should fall in love with children. There's an innocence, and yet there's a knowingness. Kids have innocence, but they're not stupid; it's all there. For a community to watch children taking ownership and having a voice is very inspiring.

Denton: When you see that joy, innocence, and commitment from children, all things that as adults we can lose, it re-inspires our connection. It helps us to remember, "No no, this is not just an intellectual thing on paper with numbers. This is about life."

INTERVIEW WITH PARTNERING ENERGY ENGINEER

Joshua Sperling, National Renewable Energy Laboratory (NREL) and Nephew to Arthur Fredric

Osnes: What inspired you to commit so much of your time and energy to this arts-based form of youth engagement on energy, climate, and resilience issues?

Sperling: First, some background that may help: my work as an "urban futures and energy-x nexus" engineer focuses on new concepts that integrate people and communities for collaborative design of integrated, high-impact solutions.

My field of study is at the nexus of energy systems, infrastructure services, and cross-sector institutions that can help improve quality of life for all, via local sustainable, healthy, resilient, and smart city and community-based strategies that draw on diverse forms of local knowledge.

I participated both to have an opportunity to work with creative young people, and to explore new forms of participatory approaches to our efforts on issues of sustainability, improved health, smart city connectivity, and building resilience across communities and cities—in ways that are inclusive of diverse populations and expertise. For fairly minimal time/energy inputs—relative to the youth, visionary individuals, and family (!) who truly co-designed this unique engagement process in multiple cities—I'm confident there'll be significant long-term returns for all involved!

Osnes: What did you gain?

Sperling: Many new perspectives, ideas and ways forward for community engagement. *Shine* is a truly wonderful example of new ways for youth to participate in their community to learn about key challenges and engage in planning ahead for lasting solutions on energy, climate, and resilience.

Osnes: What do you think the youth gained?

Sperling: The youth gained new friends, mentors, and a voice to engage in important societal challenges.

Osnes: What do you think audiences gained by being a part of this performance experience?

Sperling: The audiences gain a new appreciation for the roles of the arts and youth in realizing change. Audiences also seemed re-energized for the important work ahead—an influx of new creativity and "out-of-the-box" thinking may now emerge as a result of this performance experience.

Osnes: What are a few most memorable moments from this experience?

Sperling: The images of youth having fun learning, and proactively engaging in a web of inter-generational, trans-disciplinary, and diverse community-based connectivity continue to stand out from the experience. Moments where these forms of interconnection occurred in planning, design of, and follow-up to the performances made these engagements so unique.

I recall seeing truly engaged audiences so delighted and "all smiles" after performances. These responses offered validation and useful feedback for these continuously evolving and flexible methods. These new approaches to community engagement will likely grow in demand and offer great starting points for community-action planning that builds on forward-looking ideas shared by/led by youth, together with engineers, planners, scientists, city and community decision-makers active in new integrated approaches to energy, climate, and resilience.

Osnes: Why are the arts and performance important to issues of energy, climate, and resilience in terms of communicating the human story, or in terms of community engagement?

Sperling: The arts can help bring more participatory and interdisciplinary, evidence-based approaches. They also help form early linkage of disciplines, skills, and diverse stakeholders for local and inclusive decision-making that is critical to avoiding/responding to past and current societal failures.

Osnes: What would you say to an organization or city to encourage the use of arts and performance for youth-engagement in resilience planning?

Sperling: The example of *Shine*—as an integration of education, research, and community engagement models—can offer any organization, city, or community new perspectives and multifaceted approaches to addressing energy, infrastructure, and resilience in the USA and globally.

MATERIALS FOR BUILDING A CURRICULUM

Adaptable for K-12 Schools, Youth Groups, and Faith Organizations, Compiled by Shira Dickler

Full Production: *Shine* can be used in its entirety to actively engage students in issues related to energy, climate, and resilience. As described in Chap. 3, *Shine* can be mounted with a group of youth and performed for an audience within one intensive day of creative immersion and continual rehearsal and preparation (see Chap. 3 Boulder National Center for Atmospheric Research, London Riverside School, and New Orleans). The entire production can also be mounted over the course of a few days (see Chap. 3 Tuba City, University of East London, New York City, and Connecticut) or a few weeks (see Chap. 3 Boulder Sustainability, Energy and Environment Complex). If using the choreography as created by Arthur Fredric (http://www.insidethegreenhouse.org/shine/shine_choreography.html) it is advisable to give yourself several rehearsals for each module depending on the aptitude and experience of the performers. A video recording of the Boulder National Center for Atmospheric Research production is available online (when the production went by the name *Sol-Her Energ-he*) that could be used for guidance if you would prefer a much easier and more accessible choreography for each module (https://www.youtube.com/watch?v=gsnbX8gLfq0).

Modules: *Shine* is also designed in sections that can be used independently to engage students in one particular issue or theme. These are intended to be viewed as independent modules to explore various themes addressed by each section. What follows is a rich description of each of these various modules. Included in the description of each is:

- Themes explored
- Discussion and research questions
- Suggested warm up activity
- Artistic activity (completes with a list of materials needed)
- Description of the movement in that section (along with a link to a video demonstrating that movement).

Youth-led Movement: If the teacher or leader does not feel entirely comfortable or able to lead the movement portion of a module, check in with the youth you are working with to see if there are youth who would be willing to facilitate that portion. In my experience, a few youth in each group have consistently emerged as movement leaders. In my estimation, it could be empowering for those youth to be entrusted with a leadership position. Through the tour of *Shine,* I encountered specific comments by students (see Chap. 3 Boulder Sustainability, Energy, and Environment Complex) and teachers (see Chap. 3 New Orleans) that the opportunity to shine in physical expression was especially appreciated by students who did not tend to excel in written or spoken forms of expression.

Context for Modules: To provide the larger context from which each specific module originated, there is a professionally produced video recording of the entire performance that is 22 minutes in length and is available at http://www.insidethegreenhouse.org/shine/ that could be shown or assigned to students. Discussion questions are provided in this chapter that could be used to facilitate a discussion after viewing the video of *Shine.*

Performance in a Day: If you are planning on mounting and performing *Shine* in a single day, here are some suggestions that may contribute to the success of your experience. Before youth arrive for the day, lay out all the supplies for the art projects that need to be accomplished, such as the capes, strips of paper for the weaving, the fossil fuel flags, cutting the tissue paper, and making the plants stacks out of rolled paper. These can be decorated during breaks and by students not rehearsing a specific portion. Some of these tasks can be done before this day if possible. Note that in London at the Riverside School, the art teacher guided several of her art classes in creating all of the needed artistic items for the production in the preceding weeks. You may want to create two sets of the strips of paper for the weaving so that you have one set for rehearsal that you leave undecorated, and one to use for the performance that is decorated. If possible, fully rehearse the actors portraying Sol and Foss before the day of production so that additional focus can be given to the ensemble. If that is not possible, try to get the scripts to those actors before so they can become familiar with their lines and their characters. In London at Riverside School, university students from the University of East London performed the leading roles and the primary students performed as the ensemble only. In New Orleans one of the teachers performed the role of the Seed Sower, which helped with the co-ordination of the movement pieces. It is also beneficial for the facilitators to familiarize themselves with all the aspects of the production and have a rehearsal prior to the day of production to determine how each part will be facilitated. See the description of Boulder, CO National Center for Atmospheric Research in Chap. 3 to read how a production-in-a-day was organized in regard to timing and other considerations. Many questions about how to facilitate *Shine* will likely be answered by reading the detailed accounts in Chap. 3 of how *Shine* was produced in different communities along the tour.

Description of Modules
Act 1
Module 1. Opening, includes the entrance song *Long Time Comin'*, entrance of the dinosaur and humanity's first fire
Themes Explored in This Section

1. **Carboniferous Period** "The Carboniferous period, part of the late Paleozoic era, takes its name from large underground coal deposits that date to it. Formed from prehistoric vegetation, the majority of these deposits are found in parts of Europe, North America, and Asia that were lush, tropically located regions during the Carboniferous. (http://www.nationalgeographic.com/science/prehistoric-world/carboniferous/)
2. **Expansion of Ancient Plants and Animals on Land** "Carboniferous coal was produced by bark-bearing trees that grew in vast lowland swamp forests. Vegetation included giant club mosses, tree ferns, great horsetails, and towering trees with strap-shaped leaves. Over millions of years, the organic deposits of this plant debris formed the world's first extensive coal deposits—coal that humans are still burning today." (http://www.nationalgeographic.com/science/prehistoric-world/carboniferous/)

"Oil and natural gas were created from organisms that lived in the water and were buried under ocean or river sediments." (https://fossil.energy.gov/education/energylessons/coal/gen_howformed.html)

3. **Photosynthesis** "Many people believe they are 'feeding' a plant when they put it in soil, water it, or place it outside in the Sun, but none of these things are considered food. Rather, plants use sunlight, water, and the gases in the air to make glucose, which is a form of sugar that plants need to survive. This process is called photosynthesis and is performed by all plants, algae, and even some microorganisms. To perform photosynthesis, plants need three things: carbon dioxide, water, and sunlight." (https://ssec. si.edu/stemvisions-blog/what-photosynthesis) Photosynthesis has a direct relationship with historical carbon levels and climate change. (http://www.columbia.edu/~vjd1/greenhouse.htm)

4. **Formation of Fossil Fuels** "Fossil energy sources, including oil, coal and natural gas, are non-renewable resources that formed when prehistoric plants and animals died and were gradually buried by layers of rock. Over millions of years, different types of fossil fuels formed—depending on what combination of organic matter was present, how long it was buried and what temperature and pressure conditions existed as time passed. Today, fossil fuel industries drill or mine for these energy sources, burn them to produce electricity, or refine them for use as fuel for heating or transportation. Over the past 20 years, nearly three-fourths of human-caused emissions came from the burning of fossil fuels." (https://energy. gov/science-innovation/energy-sources/fossil)

Discussion/Research Questions

1. How do plants and animals leave behind fossils?
2. How do plants get their energy to grow?
3. How do you think "fossil fuels" got that name?
4. Why does the burning of fossil fuels release carbon?

Suggested Warm up Activity
Running Through Mud

- **Objective:** Warm the group up and generate energy

  Relevance of Activity to Theme: guides students in experiencing various ways that life moves in different environments

- **Activity:** Ask the group to move around the room using the entire space. Give instructions to the group that will change the way they are moving:

  – Walk quickly
  – Walk slowly
  – Walk on the heels of your feet
  – Hop on one leg
  – Walk as if barefoot on sharp rocks
  – Move forward as though you are in the water
  – Walk as if you were moving through mud
  – Walk like you are on ice

Artistic Activities

Costumes for Ancient Plants and Animals Create body "capes" of ancient plants and animals—a piece of Tyvek that can reach from the student's shoulders to the floor in length and that extends in width approximately from one elbow to the other (about 36"/45"), cut the corners of the rectangle to round the four corners. Cut arm holes near the top so the student can put their arms through and wear the cape on their back. Have students decorate the backs with the design of an ancient plant or animal that they research and choose to be in the performance. In the accompanying video (http://www.insidethegreenhouse.org/shine/) green full-body Lycra suits were worn by the plants and only the animals wore the capes.

Materials Needed

- Pieces of Tyvek (about 36"/45") for each student to make a "cape" of their ancient plant or animal (Tyvek is better than paper since it will not likely rip when the students move in their capes)
- Longer pieces of Tyvek to cut out and design dinosaur (can represent a dinosaur in any other way too)
- Multiple wide-tip colored water-based markers
- Scissors
- Cut up black tissue paper into small pieces (to serve as the black carbon released from humanity's first fire)

*Also needed for this section is a huge piece of brown cloth that will be spread over the ancient plants and animals once they die, approximately 20/24 feet.

Properties: Students can make a papier-mâché head for the dinosaur and green gloves to indicate the arms of the dinosaur. Multiple people can bunch together to perform a single dinosaur. Students can be allowed creative freedom to decide how they would like to create the dinosaur that will enter the stage and then exit into extinction. In the trailer for *Shine* (http://www.insidethegreenhouse.org/shine/) there are two examples of how students decided to perform the dinosaur, one with a papier-mâché head and one made out of green butcher paper operated by three-people. In the trailer is also a demonstration of the first fire by humans and the carbon being released.

Description of Movement in This Section

- First, students will create a physical representation of the historical stages explored in this scene, including: ancient plants and animals exploring the planet/photosynthesis/animals eating plants/animals dying and getting covered up by dirt and mud and sand/animals and plants being compressed to form fossil fuels (under brown cloth). While under the brown cloth, students will remove their plant or animal costume and roll out from under the cloth, leaving the "fossil" beneath the cloth. On cue, several students will act as dinosaurs walking across stage. Then, two "humans" will come out and act out lighting a fire with sticks. As the fire is lit, carbon is released as small pieces of black tissue paper into the air.

Module 2. *Harvest Song*/Foss Folks
Themes Explored in This Section

1. **Agriculture** "The history of agriculture is the story of humankind's development and cultivation of processes for producing food, feed, fiber, fuel, and other goods by the systematic raising of plants and animals. Prior to the development of plant cultivation, human beings were hunters and gatherers. The knowledge and skill of learning to care for the soil and growth of plants advanced the development of human society, allowing clans and tribes to stay in one location generation after generation. Archaeological

evidence indicates that such developments occurred 10,000 or more years ago. (http://www.newworldencyclopedia.org/entry/History_of_agriculture)

2. **Settlement of Cities Due (in part) to Grain Storage** "The concept of the 'urban revolution', first identified by V. Gordon Childe (1892–1957 CE), describes a series of social changes that brought about the development of the earliest cities and states. These changes (such as the origin of social classes and the production of an agricultural surplus) provided the social context for the earliest cities. Once class-structured state societies took hold in a region, individual cities rose and fell in response to a variety of forces." (http://www.ancient.eu/city/)

3. **Rural vs. Urban Attitudes towards the Earth/Environment** Although there are many factors influencing attitudes towards the Earth and the environment, this section portrays an exaggerated attitude on the part of Foss who disregards the rural harvest grown by his sister's energy because he prefers the fast-paced life of an urban lifestyle largely powered by his energy source, fossil fuels. This behavior is influenced as much by sibling rivalry as it is rural/urban attitudes, yet there are findings that suggest that those who live in the "rural context present more attitudes of environmental responsibility and greater consistency on expressing behavioral intentions compatible with the protection of the environment." (https://www.researchgate.net/publication/232522875_Rural-Urban_Differences_in_Environmental_Concern_Attitudes_and_Actions)

This section is not intended to assert superiority of rural over urban or vice versa, but rather to dramatize both the tension that can result from different attitudes, and the harmony that can be achieved.

Discussion/Research Questions

1. What grain allowed for your city to form? Corn, wheat, or rice?
2. What are some of the characteristics of communal-based agricultural practices?
3. How does the introduction of fossil fuels change these?
4. Do you live in a rural or an urban community? How does where you live influence your attitude towards the environment?

Suggested Warm up Activity
1 by 2 by Bradford

- **Objective**: Fostering concentration and working together

  Relevance of Activity to Section: demonstrates aspects of human development, how one behavior can be replaced by another in a human agreement

- **Activity:** Have everyone partner up. Start by telling each pair to count to three, but by alternating numbers (person A says 1, person B says 2, A says 3, B says 1, A says 2 and so on). After one minute of that, tell the groups to continue doing this, but replace 3 with a sound. Let all the groups practice that for one minute, and then tell them they now need to replace 1 with a movement. After one minute or so of practicing that, tell each group that they now have to replace 2 with a movement and sound. Let the pairs continue for another minute, and encourage them to experiment with changing the tempo, volume, and energy levels.

Artistic Activities

Properties: Harvesters create stalks of plants out of rolled newspaper by taking four half pieces of a full-sized newspaper. Begin rolling one piece from the narrow end. When half way done rolling the first piece of newspaper, slip the end of another piece into the roll and include it. Then when half way done rolling the second, slip in the end of a third piece into the roll and include it. Repeat for the fourth piece in the same way. Once completed, gently place a rubber band around one end of the rolled-up papers, careful not to make it so tight that it crumbles the paper. Take a strong pair of scissors and cut about three or four vertical slits into the other end, extending the cuts to about half the length of the roll. In performance when the Harvesters want to represent having the plant grow, they can reach into the center of the roll and pull out a center strip, which should result in a flowering of the plant. Take care not to pull too far so that the entire structure falls apart. Rehearse making these and pulling them out. If more color is desired, a colorful piece of tissue paper can be added over the second, third, and forth piece of newspaper and rolled in.

Costumes: Foss followers—Create sashes to wear across their chests that identify themselves as being aligned with Foss. Youth participants can be encouraged to design sashes in whatever way they envision to communicate their allegiance with a fossil-fuel-based lifestyle. Once designed, these can be fitted to each participant by stapling the sash so it hangs securely across the chest. Foss followers can don sunglasses to accentuate their look.

Materials Needed

- Tissue paper (yellow, orange, and green)
- Newspaper
- Rubber bands
- Scissors
- 4" × 40" strip of Tyvek for each "Foss follower sashes"
- Stapler
- Wooden bowl (to hold tissue paper "seeds")

Description of Movement in This Section

Youth participants will demonstrate the following activities based on the group they are a part of, Harvesters or the Foss followers. Harvesters dramatize tilling soil, planting (using tissue paper seeds), and having their plant stalks grow. Foss followers knock these stalks out of their way as they enter and follow Foss's lead in the dance that follows. Both groups in the second part of the song demonstrate how the two approaches to life can both clash and co-exist with vitality. (Consult the video labeled *Harvest* at http://www.insidethegreenhouse.org/shine/shine_choreography.html for a bird's-eye view of this movement)

Module 3. *Weaving*

Themes Explored in This Section

1. **Fabric of Community** Humanity, wanting to protect itself from the weather, began weaving cloth during Neolithic times. Soon, they introduced technology to help them do it more effectively, eventually using fossil fuels to fuel weaving machines. (http://www.historyworld.net/wrldhis/PlainTextHistories.asp?ParagraphID=cas)

2. **Early Machines Using Energy** Textiles are associated with the very beginning of the Industrial Revolution—the social shifts

that followed the development of weaving can elucidate the many changes occurring in human society during this rapidly changing time. Weaving went from being a family activity that used human power and a loom, to a skilled craft, to a mechanized process done in factories. (http://www.weavedesign.eu/weaving-history)

3. **How Communities Come Together and Distinguish Themselves** Design of fabric can be associated with different communities, "with patterns produced in different parts of the world showing distinctive local features" (https://www.britannica.com/technology/textile). Fabric produced through weaving is just one of many aspects of a community that can make it unique and distinguishable from other communities.

Discussion/Research Questions

1. What are some characteristics of the community you live in? What brings people together?
2. How does the weaving process make fabric stronger?
3. What distinguishes your community for other communities?

Suggested Warm up Activity
Instant Images

- **Objectives:** Communication, tackling an issue, and building a discussion

    Relevance of Activity to Section: Focuses on visual and physical representations of issues

- **Activity:** Decide on a theme to work on with the group related to the lesson unit. Everyone stands in a circle facing outwards. The leader shouts out a key word that is related to the issue, counts to three, and then claps. On the clap, the players turn into the circle and make frozen images of the word using their bodies. After giving everyone a few minutes to look at each other's image, ask for volunteers to talk about their images and why they choose them. This helps facilitate discussion on a certain issue and lets the players express themselves through their bodies.

Artistic Activities:
Properties Students will decorate long pieces of paper with images that represent their city, making up the "fabric of their communities" when woven together, such as monuments, schools, nationalities, religions, sports, bodies of water, businesses, popular pastimes, favorite foods, types of transportation, flags.

Materials Needed

- Set of 8 20' × 1' strips of paper in various colors
- Multiple wide-tip colored water-based markers (avoid tempura paint as it will make the paper strips difficult to roll again after being decorated)

Description of Movement in this Section: Performers will create a human loom by doing the following:

- Arrange 16 performers into a square, with 4 along each side. The students on the top and left sides (8 in total) will each have a roll of paper. If you do not have enough performers, you can ask people in the audience to volunteer to do the more passive roles in this weaving process—have these volunteers simply be the ones to receive the roll of paper, rather than the ones who walk through to hand over the end of the roll of paper.
- Performers holding rolls will weave them together, alternating between the top and left sides. Performers will hand the end of the rolls to their partners on the opposite sides.
- Once the weaving is complete, students on the bottom will kneel down and slant the fabric at an angle so allow spectators to see the finished product.

(Consult the video labeled *Weaving* at http://www.insidethegreenhouse.org/shine/shine_choreography.html for a bird's-eye view of this movement.)

Module 4. *Progress*
Themes Explored in This Section

1. **Mining Fossil Fuels** Since fossil fuels are formed by being compressed by rock and mud and sand, they are naturally found underground and need to be mined from the ground to be used. There

are many types of mining used in response to how deep the fossil fuels are and where they are found. (http://techalive.mtu.edu/meec/module19/Page1.htm)

2. **The Industrial Revolution** "The Industrial Revolution, which took place from the 18th to 19th centuries, was a period during which predominantly agrarian, rural societies in Europe and America became industrial and urban. Prior to the Industrial Revolution, which began in Britain in the late 1700s, manufacturing was often done in people's homes, using hand tools or basic machines. Industrialization marked a shift to powered, special-purpose machinery, factories and mass production. The iron and textile industries, along with the development of the steam engine, played central roles in the Industrial Revolution, which also saw improved systems of transportation, communication and banking. While industrialization brought about an increased volume and variety of manufactured goods and an improved standard of living for some, it also resulted in often grim employment and living conditions for the poor and working classes." (http://www.history.com/topics/industrial-revolution)

3. **Use of Fossil Fuels to Power Machines** "The new form of mineral-intensive economy pioneered in Britain during the late 1700s, and imitated in the USA and beyond in the centuries since, encountered no such limits. Instead of drawing upon limited flows of energy through surface ecosystems, mineral-intensive economies accessed much greater supplies of energy by extracting ancient stocks of energy from beneath the earth in the form of coal, petroleum, and natural gas. Fossil fuels essentially enabled Americans to harness the power of ancient suns. Coal-powered technologies magnified the strength, stamina, and precision of American workers, making the US labor force the most productive in the world." (http://teachinghistory.org/history-content/beyond-the-textbook/23923)

4. **Disruption of Carbon Cycle through Carbon Emissions from Fossil Fuel Use** "Without human interference, the carbon in fossil fuels would leak slowly into the atmosphere through volcanic activity over millions of years in the slow carbon cycle. By burning coal, oil, and natural gas, we accelerate the process, releasing vast amounts of carbon (carbon that took millions of years to accumulate) into the atmosphere every year. By doing so, we

move the carbon from the slow cycle to the fast cycle. In 2009, humans released about 8.4 billion tons of carbon into the atmosphere by burning fossil fuel." (https://earthobservatory.nasa.gov/Features/CarbonCycle/page4.php)

5. **Who is Impacted the Most by Climate Change?** "Many women around the world must adapt their lives to a changing climate. Increases in extreme weather conditions—droughts, storms, and floods—are already altering economies, economic development, and patterns of human migration, and are likely to be among the biggest global health threats this century. Everyone will be affected by these changes, but not equally. Vulnerability to climate change will be determined by a community's or individual's ability to adapt. Studies have shown that women disproportionately suffer the impacts of disasters, severe weather events, and climate change because of cultural norms and the inequitable distribution of roles, resources, and power, especially in developing countries." (http://www.prb.org/Publications/Articles/2012/women-vulnerable-climate-change.aspx)

Discussion/Research Questions

1. What are some ways that you use fossil fuels in your life? What are the benefits you receive from this?
2. Are any fossil fuels mined in or near your community? What extraction method is used?
3. What type of industry is powered by fossil fuels in your area?
4. Can you imagine what it would be like to survive in your area without access to fossil fuels?
5. How does the release of carbon from this industry impact climate?
6. Who in your community is most vulnerable to a changing climate and why?

Suggested Warm up Activity
Machine

- **Objective:** have a physical experience of industry and power

  Relevance of Activity to Section: provides a physical experience of how machines and fossil fuels help to power a city

- **Activity**: One person comes to the center of the circle and repeats a mechanical sound and movement. One at a time, everyone else joins in with their own sound and movement in such a way that each movement is interrelated to one other person, thus making a human machine. Then create another machine by asking them to think of themselves as the different parts of their city that help it function, such as trash removal, police, schools, hospitals, and water treatment. With that in mind, create another machine meant to represent their city.

Artistic Activity
Property

- Students will decorate large black banners as flags that represent the many ways in which their city uses fossil fuels (example: heating homes and businesses, power plants, transportation, street lights, buses). Both sides of the flags will be decorated and taped to 5' wooden poles.

Materials Needed

- 4–9 pieces of black Tyvek cut in rectangles 36"/45"
- Multiple wide-tip colored water-based markers
- 4–9 5' wooden poles
- Black duct tape (for affixing banners to poles)

Description of Movement in this Section: During this section, the Foss Follower (performers who are not playing the role of weavers) will do the following:

- Hold flags and circle around the weavers (who are still holding the fabric of community).
- Portray the march of progress with strength and determination, miming the digging up of the fossil fuels.
- Once the storm starts, performers circle the weavers as though caught up by the wind of the storm, allowing the flags to represent the strong winds, all the while the movement of the performers becomes more erratic.
- At the last "clang" of the storm, two or three performers should rip through the fabric of community destroying it.

- One of the weavers is hurt and falls to the ground into Foss's arms just as the storm stops. Foss looks up to Sol and asks "what now?" to complete the final tableau of Act 1.

(Consult the video labeled *Progress* at http://www.insidethegreenhouse.org/shine/shine_choreography.html for a bird's-eye view of this movement.)

Act 2

Module 5. Performance of Youth-Authored Solutions Through Skits and Celebratory Song

This module will have a different format than sections from Act 1 since this section is to be authored by youth participants. Rehearsing and performing Act 1 is designed to prepare and inspire youth participants to author solutions for Act 2. Alternately, a teacher who is only using this module may choose to simply have students watch the video of Act 1 (full performance available at http://www.insidethegreenhouse.org/shine/)

and then guide their youth in creating their own skits that express local solutions that they identify. Skits could be performed theatrically live or created as short videos. What follows are some suggestions for guiding youth in identifying solutions as a group, in creating two different types of skits based on those solutions, and considerations for effective communication of solutions. Many more approaches to creating skits are included within the descriptions of each production of *Shine* along its tour in Chap. 3. Many more approaches can be developed beyond these. Discuss options with your performers—matching the method to the needs, personality, talents, and preferences of the youth with whom you are working.

Guiding Youth in Identifying Solutions in Groups

To ensure that each participant's idea within each group will be heard and considered, ask each group to sit in a circle, close their eyes, and each silently think of a solution to a local challenge regarding climate, energy, or resilience. Ask each person to share their idea in just one sentence with their group, noting that the group will be choosing just one idea, but that they don't have to decide yet. After each has shared, ask the groups to consider if any common themes emerged or if they see a natural way of synthesizing the solutions that were shared. Ask them to consider which solution might lend itself best to being acted out for the audience. Give groups about two to five minutes to come to consensus on a solution they would like to enact for their skit.

Enacting a Skit as a Narrated Statue (process takes about 20–30 minutes depending on how many youth are in the group)
This is a method for creating a skit that can be achieved in a limited amount of time. Divide students into groups of four. Each group will be using their bodies to create a statue that conveys their solutions in an active and interesting way. One person in the group will stand to the side of the statue to narrate for the audience what solution is being communicated by this statue. Give each group about four minutes to create their statue or movement, and to decide who will narrate and what that spokesperson will say. Each group can perform their solution for the rest of the groups to receive positive feedback and constructive suggestions for expressing their solution even more clearly. Provide another two minutes for groups to reconvene to integrate any suggestions they received. For descriptions of youth-authored solutions using this method, read the Descriptions of Youth-Authored Solutions from New Orleans, Connecticut, and New York City in Chap. 3.

Generating a Skit Using Image Theatre (Process Takes About 40–70 Minutes Depending on How Many Youth Are in the Group)
Once each group has decided on an issue for their skit, ask them to create three distinct images: (1) an image of the problem, (2) an image of the solution to the problem, and (3) the transitional image, or an image of the action that got them from the problem to the solution. By "image" I mean a frozen scene made up of their bodies that physically communicates each prompt. Ask them to portray a specific manifestation of the problem. For example, if the problem is homelessness, the image of the problem might be a single old woman sitting on the sidewalk reaching up for spare change as two other people walk by her with their chins up and their gaze avoiding her. The solution might be an image of this old woman in a co-operative living residence making a meal with other residents. The transition from the problem to the solution might be an image of neighborhood residents in the office of their government representative advocating for housing for the homeless.

Once each group creates their three images, have them take turns sharing these with the others, one group at a time. First, they show the image of the problem, then the solution, and finally the transition. They are not allowed to use any words when presenting these nor are they allowed to announce what their issue is. Once each group is done, ask the others to reflect to the group what they saw. This gives each group a chance to hear what was communicated clearly and what might need more description or clarification. Then give them time to create a skit based on the same issue that is about one to two minutes in length. Urge them to be playful with the creation of their skits, not to over-think them, but, rather, to get on their feet and actively work through the creative process. To support this, only give them ten minutes to create their skits as a group, after which time they each will share their skit to receive positive feedback or constructive suggestions for improvement. Beginning with the Image Theatre exercise can assist in emphasizing the embodied aspect of their communication. For descriptions of youth-authored solutions using this method, read the Descriptions of Youth-Authored Solutions from University of East London in Chap. 3.

Considerations for Effective Skits:

Research shows that when communicating solutions to climate and energy related issues, it is useful to:

- Keep it local—framing at the city/community level
- Appeal to people's already held values
- Focus on a single issue
- Emphasize the positive
- Identify co-benefits to climate and energy solutions
- Frame the solution as an opportunity[1]

Final Song: *Shine*

Importance of Celebrating at the End in Song and Dance:

It is a tradition in cultures throughout the world to end public gatherings with an inspirational song and dance that ensures the sustainability of the energy and commitment necessary to follow through with the issues addressed at the gathering. If you leave humming the final tune, that may help you carry the spirit of commitment with you into your daily life. It may infect you with the inspiration built into the event purposefully by its organizers. Participating in the final number can give you an experience of connection and joy that allows a person to feel the value of their community, which in turn, will hopefully strengthen the resolve to act on its behalf. Shared cultural expression unites us, allows us to feel who we are as a community, and communicates who we are beyond our borders.

Choreographing *Shine*

After rehearsing and performing the established choreography of Act 1, the choreography for this song can be handed over to the ensemble of youth performers. This is an example of scaffolding the lesson of how movement can be used to contribute to the expression of an issue or theme, first by teaching the students choreography of Act 1 and second by giving them the independence to choreograph the final song of Act 2 themselves. The lyrics of the song *Shine* lend themselves easily to simple yet expressive movements. Encourage students to decide upon movements that are accessible to everyone in the group and that are not too complex so that joy and release can be experienced while performing. (Consult the video labeled *Shine* at http://www.insidethegreenhouse.org/shine/shine_choreography.html for a bird's-eye view one version of this movement)

ADDITIONAL DISCUSSION QUESTIONS

There is a professionally produced video recording of the entire performance of *Shine* that is 22 minutes in length and is available at http://www.insidethegreenhouse.org/shine/. The following questions could be used to facilitate a discussion after viewing the video of *Shine* or for the use with one of the modules described above.

1. Science-Based Questions

- How do plants and animals leave behind fossils?
- How do plants get their energy to grow?
- How do you think "fossil fuels" got that name?
- What are some of the characteristics of communal-based agricultural practices? How does the introduction of fossil fuels change these?
- What is the carbon cycle? How can it be "disrupted," as discussed in the show?

2. Literature-Based Questions

- Why is "Foss" considered the brother of "Sol"?
- What does Sol think of carbon emissions at the beginning of *Shine*? How does her perspective change throughout the script?
- How do the "Foss Followers" feel about the Harvesters and their lifestyle and vice versa?
- Why does Foss's idea for progress make Sol so concerned?

- How is metaphor used in this play? Is this an example of literary metaphor or dramatic metaphor?
- What is the effect of having anthropomorphized characters? How does this impact your understanding of the scientific concepts presented?

Sample Exercises and Games

Below are descriptions of a few performance exercise and games referenced throughout this book, along with others that may be useful to prepare youth for being expressive. Many other such exercises, activities, and games exist and can be found in applied theatre texts,[2] as well as texts specifically for participants who have different physical and mental abilities[3] (see Fig. 2.1).

Shaking out the Tension: Quickly count from one to eight, shaking one hand, the other hand, one foot, and then the other foot. Repeat while counting to four, then to two, then to one.

Fig. 2.1 Students playing a game before rehearsing *Shine*. Photo by Conner Callahan

Name and Gesture: Standing in a circle, have one person say their name while doing a gesture or movement that expresses how they are feeling at that moment. Immediately have everyone repeat the name and the gesture altogether. Ask the next person to say their name while doing a gesture or movement that expresses how they are feeling, after which everyone repeats this. Go around the circle until everyone has done this.

Zip Zap Grr: Standing in a circle, ask everyone to press their palms together in front of them. Pass the energy around the circle in one direction by moving your hands towards the person next to them and saying "Zip," then the next person does the same saying "Zap," and so on, repeating this pattern and the words "Zip" and "Zap." As the leader, have the group practice passing the energy, moving one way completely around the circle, and then the other. Then ask everyone to practice putting their hands up like claws and growling a "Grr." While passing the Zip Zap energy around the circle, if the person you pass it to turns towards you with a "Grr," you will become frightened and pass the energy in the opposite direction. Anyone can change the direction of the energy with a "Grr" whenever they like. A further option is to do the exercise and if anyone makes a mistake in the game, s/he steps out of the circle until it is down to just two winners.

Rainforest: Standing in a circle, explain that as the leader you will start an action. The person to your right will repeat that action, and then the person to that person's right and so on. Once this action reaches the leader, the leader will introduce a new action. In this way, the leader will "send around" various sounds made with the body. The actions that the leader send around in turn include: rubbing hands together, snapping, clapping, stomping, clapping, snapping, and rubbing hands together, and then nothing. The goal is for this to replicate the sound of a rainforest in which a gentle rain comes, builds to a heavy storm (represented by the stomping), weakens, and then subsides.

Water Bottle in the Middle: Standing in a circle, put an empty water bottle in the middle on the floor. Invite anyone in the circle to come forward and interact with the water bottle as though it is something else (a football, a baby, a candle,) until someone can guess what the water bottle represents. Once it is guessed, the person puts down the water bottle in the middle and returns to the circle. Someone else can continue the challenge, each person trying to evoke a different item with the water bottle.

Count to Ten: Standing in a circle, have the group count to ten together with only one person speaking at a time. If two people say a same number at the same time, they have to start over again with the number one. The goal is to count to ten as a group without anyone speaking over another person. Do not allow the group to take turns counting in order (or it becomes too easy.) If the group masters this challenge, see how high they can count using these rules.

Morph-Ball: Standing in a circle, the leader holds an imaginary ball. Mold and shape that ball with your hands to establish its size, weight— even bounciness—and then throw it to another person, who catches it in a way that acknowledges that ball's properties. The second person then morphs the ball into another size and weight, and throws it to a third person, and the pattern continues.

Telephone with a Twist Start with everyone in a line. The person on one end of the line starts the telephone. Instead of passing down a phrase, they pass down an explanation of a movement (such as "hop on one foot in a circle while patting your head"). The line continues to pass down the instruction until it gets to the end. The last person in the line then has to do the movement that was instructed, and see how close they were to the original.

Rock, Paper, Scissors Championship: (Ensure you have an even number of participants for this game. The leader can participate if needed.) Ask everyone to get with a partner. Review rules for playing Rock, Paper, Scissors. Have everyone practice together a few times to be sure everyone is in rhythm with the group. Begin the grand competition by having each group of partners play. Whoever loses gets behind their partner to cheer that person on. Whoever wins, raises their hands above their head and should look for another winner to play. Repeat this until it is down to two final champions (each with many people behind them cheering them on,) until one person wins.

Machine: One person comes to the center of the circle and repeats a mechanical sound and movement. One at a time, everyone else joins in with their own sound and movement in such a way that each movement is interrelated to one other person, thus making a human machine. Then create another machine by asking them to think of themselves as the different parts of their city that help it function, such as trash removal, police, schools, hospitals, and water treatment. With that in mind, create another machine meant to represent their city.

Resilience in Motion: Ask everyone to stand in one long straight line. A person at one of the ends is designated the lead person and turns to face the rest of the people in the line. Everyone in the line faces the lead person. The lead person reaches their hand across and grabs the next person's outstretched hand, and each pulls past the other. Each person uses the other hand to reach across to grab the next person's hand, like an old fashioned square dance. Once a person gets to the end of the line, they turn back and wait to be offered a hand and be reintegrated into the line. Everyone in the line continues repeating this pattern. The goal is for the line to be like one big organism moving together. For an added challenge, try this same movement pattern in a circle with an even number of pairs.

NOTES

1. Ezra Markowitz, Caroline Hodge, and Gabriel Harp, "Connecting on Climate: A Guide to Effective Climate Change Communication" (New York: Center for Research on Environmental Decisions, Columbia University, 2014).
2. Clarke Baim and Sally Brookes, *Geese Theatre Handbook: Drama with Offenders and People at Risk* (Hook: Waterside Press, 2002); Philip Taylor, *Applied Theatre: Creating Transformative Encounters in the Community* (Portsmouth, NH: Heinemann Drama, 2003); Augusto Boal, *Games for Actors and Non-Actors* (New York: Routledge, 2002).
3. Petra Kuppers, *Studying Disability Arts and Culture: An Introduction* (New York: Palgrave, 2014).

REFERENCES

Baim, Clarke, and Sally Brookes. *Geese Theatre Handbook: Drama with Offenders and People at Risk*. Hook: Waterside Press, 2002.

Boal, Augusto. *Games for Actors and Non-Actors*. New York: Routledge, 2002.

Kuppers, Petra. *Studying Disability Arts and Culture: An Introduction*. New York: Palgrave, 2014.

Markowitz, Ezra, Caroline Hodge, and Gabriel Harp. "Connecting on Climate: A Guide to Effective Climate Change Communication." New York: Center for Research on Environmental Decisions, Columbia University, 2014.

Taylor, Philip. *Applied Theatre: Creating Transformative Encounters in the Community*. Portsmouth, NH: Heinemann Drama, 2003.

CHAPTER 3

Outcomes from the International Tour of *Shine*

Abstract This chapter describes how *Shine* was uniquely mounted in host communities during its year-long tour. *Shine* reached four different nations where English is used: Navajo Nation (a semi-autonomous Native American territory in the USA), four cities within the USA, the UK, and South Africa. This is the most substantial portion of the book and contains the heart of what was experienced and learned from this entire project. For each performance within a different community, the following is provided: a short description, a detailed described of the process and the resulting performance, youth-authored solutions, lessons learned and recommendations, and feedback from performers and/or audience members. The intention behind the tour was to learn from the diverse approaches by host communities and youth around the world.

Keywords 100 Resilient Cities · Youth engagement · Applied theatre Climate change · Energy · Resilience · Partnerships · USA · Navajo Nation · United Kingdom · South Africa

This chapter describes how *Shine* was uniquely mounted in each host community during its year-long tour. *Shine* reached four different nations where English is a primary language: Navajo Nation (a semi-autonomous Native American territory in the USA), four cities within the USA, the UK, and South Africa. The first iteration of the performance was premiered in Tuba City, Arizona within the Navajo Nation

© The Author(s) 2017
B. Osnes, *Performance for Resilience*,
https://doi.org/10.1007/978-3-319-67289-2_3

in the Southwest of the USA through a week-long artist-in-residency at Tuba City High School. Since *Shine* was developed at the University of Colorado (CU), the city of Boulder served as its primary testing ground. In Boulder, *Shine* was mounted in a national laboratory for climate science; a complex for climate science at CU with a local middle school; and as a week-long science summer camp for youth. In New York City, *Shine* was the culminating plenary presentation for the Urban Thinkers Campus, *The City We Need* conference. This performance helped ensure the inclusion of youth voices in preparation for the October 2016 Habitat III, the United Nations Conference on Housing and Sustainable Urban Development. In London, it was performed at the conclusion of a week-long artist-in-residency at the University of East London and, the following week, with the Year 7 class of Riverside School in East London. In New Orleans, we mounted *Shine* at a private Catholic grade school—having an opportunity to address both faith and environmental stewardship in the same conversation. In Malope, South Africa, *Shine* was mounted in a rural primary school, using only the final song and skits. In Connecticut, *Shine* was organized by the show's choreographer, Arthur Fredric, with a multi-city group of performers, which allowed us to produce a professional video recording of *Shine* (https:// vimeo.com/194833723). I travelled to each location to facilitate local youth participating in these performances, always in collaboration with a local host. The intention behind the tour was to learn from: the diverse approaches to the material; the various community hosts; and from youth around the world.

The initial name for the performance was *Sol-her Energ-he*, since much of my own performance-based research and creative work has focused on the gendered nature of energy development, and how our current methods for practices, policies, and industry disproportionally impact women negatively.[1] This was most obviously conveyed through the script by having Sol portrayed by a female, Foss by a male, and by identifying the harvester who gets hurt by the climate-induced destruction at the end of Act One as a woman. What I learned less than halfway through the tour was that, although the title was clever (*Sol-her Energ-he* sounds like 'solar energy' when spoken aloud), the combination of climate as a subject plus a focus on gender had the potential to keep audiences away in droves! A title that was lively and clear of bias, such as *Shine*, would likely be more inviting to a wider more varied audience and could more accurately represent the spirit of the offering. Taking a cue from the last song in the show, *Shine* seemed a perfect expression of what youth do in this performance: they shine a light on issues and solutions, while displaying their remarkable talent.

TUBA CITY, ARIZONA, NAVAJO NATION

Where: Tuba City, Arizona USA within the Diné Bikéyah or Navajo Nation
When: March 23–27, 2015
Community Partner: Tuba City High School with the Language Arts/ Theatre teacher Gerald Vandever and Vice Principal Rubin Ruiz; and Colleen Biakeddy and Adrian Manygoats of the Navajo Women's Energy Project
Note: Tuba City High School services both the Navajo and Hopi Nations (two sovereign indigenous nations within the Southwestern part of the USA.) The Navajo Nation—the largest indigenous nation in the USA—is in the states of New Mexico, Arizona, and Utah. The Hopi Nation is a smaller area within the Navajo Nation. Although Tuba City High School is on Navajo land, it is within a mile to the Hopi reservation, so the participating students in this performance included Navajo, Hopi, and other non-indigenous students (Fig. 3.1).
Short Description: This first performance of *Shine* was the result of a week-long residence in a high school within the Navajo Nation with a lively group of CU college students co-facilitating three different classes of students. A popular local teacher with full support from the administration supported a fruitful week that culminated in an all-school assembly. CU students learned much from this cross-cultural experience in being authentic allies to indigenous concerns in the quest for a bright energy future.
Description: You may wonder how it came to pass that the captain of the Tuba City High School football team danced in an all-school assembly performance of *Shine*. This had a lot to do with the fact that Gerald Vandever, our primary host, was then the coach of the school's football team, led the weightlifting club, and has a theatre degree from Arizona State University. He is a tall, imposing man whose warm smile can set an entire classroom at ease. He is also a tireless advocate for his students, both inside and outside the classroom. As a Diné (Navajo) man, not only does he someday aspire to serve as the president of the

Fig. 3.1 Tuba City High School students and the Hopi Language Arts teacher dancing in the final number *Shine*. Photo by Conner Callahan

Navajo Nation, he plans to gain votes by being as authentic a partner to youth as he possibly can be. So, when Vandever hosted us in his class, not only did students in many other classes beg their teachers to join our class, but he easily enlisted the spirited participation of members of the football team. All of this made him an ideal partner for our first mounting of *Shine* through a week-long artistic residency at Tuba City High School to engage students in exploring energy and climate issues.

The genesis of *Shine* originates in the work I did in partnership with Adrian Manygoats and numerous other Navajo women in creating the Navajo Women's Energy Project (NWEP). One of the recommendations that emerged from NWEP was to create a way for young people to gain education and awareness around energy issues. Much of the current Navajo economy is based on extraction of energy resources—namely coal and uranium—and energy generated from those resources. Jobs and revenue from energy production are thus linked to devastating environmental and health damages, such as contamination of water tables, high cancer rates, and scarring of the land from strip mining. According to the website for Eagle Energy, about 38% of households in the Navajo Nation lack electricity, and "despite requests for modern, grid-based power, the remote location of many Navajo households makes electricity extremely expensive, forcing many people to rely on wood and kerosene for energy."[2] Many organizations such as Black Mesa Water Coalition and the Grand Canyon Trust are attempting to generate the political will to transition to a clean energy economy. Since the NWEP was formed using theatre-based methods for women to deeply consider energy issues, it seemed appropriate to use theatre as a method for youth engagement on the same issue.[3] A few elderly women in the community were able to attend the performance and were deeply gratified to witness such spirited youth expression.

With me on this week-long residency were four CU students and two members of my family, my niece Claire Hackett (who later served as the host to *Shine* in Malope, South Africa,) and my middle-school aged daughter, Lerato. We were told by Vandever to expect about 25 students in each of our sessions, but in reality, we had anywhere from 50–75, as word spread among students throughout the week. Each day we taught three classes that lasted 75 minutes each. We assigned one song to each class, with the goal that they would create the necessary props for that song and would perform it for the all-school assembly at the end of the week. We had all three classes learn the final celebratory song to perform together. The lead roles of Sol and the Seed Sower were performed by CU college students and Foss was performed by Vandever. Instead of having a script with lines—as included in this book—I stood to the side and narrated the action that was enacted by the cast "on stage." Sarah Johnson, a CU Ph.D. student, had worked with me in creating this script.

On the first day, we gathered the first class into one large standing circle and did a few of the warm up exercises listed in Chap. 2—"Shaking out Tension and Name and Gesture." The class then learned to perform the *Harvest* dance. They also learned how to roll the recycled newspapers to create newspaper trees that could be pulled out to simulate the growing of the harvest plants. To relate the theme of agriculture represented in the play, we discussed what kind of plants were grown in their areas that sustained life, and who grew them in their families or their community.

We did the same warm up exercises with the second class, before teaching them to perform the *Weaving* song. Next, we asked them to say a word or short phrase about community, one at a time around the circle. Contributions included: family, food, Hopi, love, compassion, peace, friendship, security, and cooking. We then invited them to use colored markers to decorate 12 long strips of paper with images that represent their community. These long strips of paper, once rolled tightly, were woven during the *Weaving* song by the students in a human loom created by a square of students with six on each of the four sides. These students bodily wove together the individual strands into the fabric of their community.

With the third class, we began also with Shaking out Tension, Name and Gesture, and a game of Zip Zap Grr. Standing in a circle, we asked each to say a word or short phrase about coal, oil and gas. Contributions included: America, pollution, grandma's stove, shiny, greasy, sticky, Black Mesa, Chevron, profit, and money. We then invited students to decorate flags to represent ways in which fossil fuels are used within their community. Students used these flags for the Progress song in which Foss's followers waved their flags, while circling the loom with the woven fabric of community. They also practiced being tossed and blown about by the storm caused by excessive use of fossil fuels that is altering the climate. At the climax, they rehearsed losing control and ripping through the fabric, injuring one of the weavers who falls into Foss's arms.

Each of the classes contributed some aspect to the choreography of the final title song *Shine* that celebrates the successful creation of solutions that bring back the Sun. Throughout this process, we all learned a lot about how to best use the theatrical performance to engage students in the issues of energy and climate. The process was incredibly invigorating and fun for all involved. We were truly impressed by how willing these high school students were to participate in both the physical activity and the making of the props used by each class in their performance. On final day of our week-long residency, students performed *Shine* as an all-school assembly for about 250 students, plus 20 children from a pre-school housed at the high school, teachers (including the Hopi Language Arts teacher shown in the photo singing and dancing in the final song), school administrators, and community members.

Highlight of Youth-Authored Solutions:

In this version of the story for *Shine*, the character of the Sun went into hiding because she was so upset by the damage wrought by climate change. Student performers invited audience members to approach the stage and share their ideas for getting her to come out again and bring back her life-giving light. One student suggested we try to recycle more to improve conditions and lessen the impact of climate change. One of the pre-school audience members encouraged everyone to say sorry to the Sun. Another student thought that using less energy for transportation could make a difference. Each student bravely walked up to the stage and spoke into a microphone that was being held by one of the cast members. Once the Sun was coaxed to return, the entire cast performed the final song *Shine*.

Lessons Learned and Recommendations:

1. Establish buy-in at a school from both teachers and the administration: One thing that significantly contributed to the success of this artistic residency and the production of *Shine*, was the partnership with the very popular teacher, Vandever, and the support of the vice principal, Ruiz. This required an onsite visit prior to the residency and many phone calls and emails with both the teacher and the administrator months before the actual residency. Having buy-in from a teacher and an administrator made for a warm reception and helped everything run smoothly. If a project is "bought into" by the administration and forced upon teachers, obstacles can arise; just as they can if a teacher brings a project into the school but doesn't have the support of the administration. We were able to supply a full audience for the students' hard work through an all-school assembly because the administration supported this show. They also made arrangements for a stage to be erected in the gymnasium, for there to be sound amplification, and for the gymnasium seating to be set up.

2. Relate Performance Elements to The Community's Culture: In a post-show conversation, one of the Navajo students said, "Because weaving is such an important part of our culture, when the weaving got ripped during the storm, people in the audience gasped." This element of the performance—the dramatic metaphor of ripping through the fabric of community as representing the destruction of culture by climate change—was especially powerful. One student noted in a post-show discussion how that part of the show really meant something to them, especially because so much environmental damage has destroyed parts of their community and the culture that relies on the environment, such as the sheep that provide the wool for the weaving and the way of life that sheep herding offers. This integration of elements of the community's culture seemed to invigorate and deepen the conversations about the meanings in the play both during rehearsal and after the performance. During the classes when students decorated the paper strips for the weaving section, several of them had drawn traditional designs used in weaving. Since almost every culture has some tradition of weaving cloth that uniquely represents their community, this correlation can be made explicit in nearly any community to engage students. Students could also be prompted to consider drawing textile designs onto the paper strips to represent their community's unique patters and designs. Similar local connections can be made when identifying how coal, oil, and gas are used in the student's community.

Feedback by CU Students

In our conversation about the experience on our long drive home from Arizona to Colorado, Arianna Orozco—an International Affairs senior at CU—was surprised how much the high school students participated and were open to creative activities. Before arriving, there was some apprehension among the CU students that high school students would be too cool to participate in this show. Conner Callahan—a senior CU Environmental Studies student—appreciated how issues around energy access and climate where explored within the students' cultural practices and beliefs and how their opinions on energy use and their ideas for the future were incorporated into activities throughout the week. Another CU student, Topaz Hooper—a Human Geography senior—felt relieved that she didn't feel as though she was overly imposing an already finished artistic creation onto another culture. Her experience was that it felt more like an offering, especially since there were opportunities for the students to add their own expression through drawing, movement, and the authorship of solutions. Given the historic legacy of the horrifically unfair treatment of indigenous people in the USA by dominant populations, these issues of imposing culture from the outside are important. One way we helped to mitigate this was to share a meal early on in our visit with an elder Navajo woman, Colleen Biakeddy, who is a part of the Navajo Women's Energy Project and a sheep herder. We enjoyed hours of conversation and learned a lot about the region and its history through her stories. We also shared meals and much conversation with our primary host, Vandever, and followed his lead in relating to the students.

Another tool we used to prepare for this artist residency was an Ally Agreement created by a group called Purpose Focused and shared with us by Don Yellowman of Forgotten People, an organization that advocates for the rehabilitation of the Former Bennett Freeze Area in Arizona. On the drive down, I gave each student a printed copy, and then upon our arrival we discussed its contents with Adrian Manygoats. The agreement is an ally support memorandum of understanding and asks non-indigenous allies to agree to various directives such as not acting out of a feeling of guilt, but rather out of a genuine interest in challenging larger oppressive power structures. It also requests that allies be aware of their privilege and that they ground themselves in their own ancestral history so as to avoid the 'wannabe syndrome,' which could undermine indigenous efforts. I think this significantly increased the CU students' ease and comfort working with mostly indigenous students. It seemed to contribute to their ability to process their experience and to deepen their understanding of the cultural context in which they were working and their positionality in regard to issues of race and privilege.

BOULDER, COLORADO, USA

National Center for Atmospheric Research

Where: Boulder, Colorado, USA
When: June 12, 2015
Community Partner: National Center for Atmospheric Research (NCAR) Diversity, Education and Outreach
Short Description: This musical-in-a-day experience in a national center for climate science was the perfect opportunity to collaborate with leading scientists in officially launching this tour. Performing at the National Center for Atmospheric Research and at the Conference on Communication and Environment contributed towards the validation of youth voice in resilience planning. Attendance by Boulder's Chief Resilience Officer resulted in youth contributions being included in Boulder's official resilience strategy.

Description

On June 12, 2015 parents drove up the winding road to drop their kids off at NCAR at 8:30 in the morning to begin an intensive all-day rehearsal that culminated in an afternoon performance at NCAR, and an evening performance at the University of Colorado. The only way this ambitious goal was accomplished was by partnering with four energetic CU students (two of whom were on the trip to the Navajo Nation previously described,) three community volunteers, the show's composer who provided live drumming, and two NCAR scientists. We had 13 youth performers ranging in age from 9 to 16 who were local students. In the weeks prior to this day, I had rehearsed with the two teenaged youth who performed the speaking roles of Sol and Foss to memorize and completely rehearse their parts. A week before June 12, the two lead actors, the four CU students, and myself met for six hours to rehearse leading the other youth through performing Act One, authoring Act Two, and facilitating the youth in choreographing the final song and dance number. I had met previously with choreographer Arthur Fredric during an artistic residency in Boulder to learn his choreography for the other dances in *Shine* (Fig. 3.2).

Paty Romero Lankao is an interdisciplinary sociologist working as a senior research scientist, and was the primary host for this mounting

Fig. 3.2 Weaving the fabric of community. Photo by Conner Callahan

of *Shine*. Our hosts at the National Center for Atmospheric Research (NCAR)—climate scientists Lankao and Joshua Sperling—were planning on just getting us settled that morning and checking in on us intermittently. However, once they felt the spirit of the room and got involved, they both stayed with us the entire day, actively participating in each portion. Throughout the day, they rehearsed the dances and movements alongside the youth, sometimes elucidating a scientific principle or idea raised by the script.

We began the morning with the Name and Gesture game, a vocal and physical warmup, and an activity to introduce the notion of city resilience through the making of a machine together (Fig. 3.3). One person came to the center and repeated a mechanical sound and movement. One at a time, everyone else joined in with their own sound and movement in such a way that each movement was interrelated to one other person, thus making a human machine. Then we asked them to think of themselves as the different parts of our city that help it function, such as trash removal, police, schools, hospitals, and water treatment. With that in mind, we created another machine meant to represent our city. While that machine was moving at full force with everyone involved, I

Fig. 3.3 Rehearsing the Machine exercise Photo by Conner Callahan

announced that a flood was coming through our city and asked them to react to that as the machine. Some people fell over, but still reached to repeat their movement and sound in conjunction to those near them. Some people helped others restore or maintain balance under the strain of the water, while a few seemed unaffected. Announcing the flood had passed, I announced that a wild fire was coming, and after that a drought. Through each disaster, different people faltered or thrived differently. Afterwards, we reflected on how each natural disaster impacted different parts of the city and the capacity of each part of the city to be resilient in the face of these shocks. We asked participants to also identify social stresses that impact our city—such as unemployment, homelessness, and racial bias—and imagine how each of those impacts various parts of our city and vulnerable populations.

Nearly all the speaking in *Shine* is done by Sol and Foss, with the rest of the ensemble bringing to life what is being spoken about. Since the spoken portions of the show were previously rehearsed, we primarily just needed to set the movement for the five songs in the show, four of which each have pre-established simple, accessible, but artistically beautiful choreography. Each of the songs was pre-recorded, and participants were encouraged to sing along as best they could. Each musical number was designed to only take about half an hour to teach to the youth. For the final title song *Shine*, we facilitated the group in choreographing the movement themselves, to support greater self-expression by the cast. We used each of the two 20-minute breaks to allow youth participants to eat a snack, but also to decorate the props we needed for the show, such as the eight colorful paper strands decorated to represent Boulder (that would be woven into the fabric of community and the five flags that needed to be decorated with images of how Boulder utilizes fossil fuels.)

Although a synopsis of the show's plot and the actual script are included in this book, what follows is a thick description of the design for the rehearsal and performance of the beginning portion of the show. The show opens in the Carboniferous Period and dramatizes when the organic deposits of plants and animals were covered by layers of earth to form the world's first extensive coal deposits. Most of the ensemble was costumed as ancient plants in green suits that cover their entire bodies with a sash of green leaves. Entering from behind the audience, they moved in vine-like patterns through the crowd, crouched and reaching with their arms as if growing. The remaining three performers wore capes with a trilobite, dragonfly, and lizard respectively painted on the back. As Sol addressed the audience, they all slithered, crawled, or flew into animated existence. The song that accompanied their entrance included the caws of ancient life and the buzz of insects along with the prerecorded words, "Long time coming, a long time a coming coming, long time coming along." The plants gathered towards the sun's warmth and danced the various parts of photosynthesis while Sol narrated their actions: soaking in her energy, taking up water through their roots (which excited their cells,) breathing in CO_2 from the atmosphere and growing. In rehearsal, performers were encouraged to create specific actions that would represent each part of photosynthesis and perform these together as an ensemble. Comic relief ensued when the ancient animals then feasted on the plants, such as a trilobite munching on a plant's outstretched arm. When Sol announced that they all eventually die, both the plants and the animals—the entire ensemble—had a delightful time each enacting the most prolonged and dramatic death they could manage. Once dead, they were covered by a huge brown cloth that represented the hundreds, sometimes thousands, of feet of mud, rock, and sand. While rehearsing this scene and when performing it, the recorded song, *Long Time a Comin'*, and then Sol's narration sufficiently guided the performers' actions. The accompaniment of a live drummer allowed us to accentuate the pulsing movements of photosynthesis in a sensuous celebration of life. Once under the brown cloth, performers were instructed to take off their green suits and capes and silently roll out the sides in their all-black street clothing. This was to represent their transformation into coal under the weight of the earth's pressure. In this way, the practical tasks of performing were intricately linked to the science being addressed. Performers were reminded to push their discarded costumes to the center under the cloth (where they could be fully hidden) before exiting so that their transformation to fossil fuels would read visually. When the cloth was then gathered up and pushed off stage by a stagehand, all record of the lush green signs of life were gone.

The rehearsal of the subsequent songs proceeded in a similar fashion, where there was some structure—provided by the songs, pre-established choreography, and narration—yet room for individual and group authorship through some of the movement and dramatic expression. Also, rehearsing each song teaches an aspect of climate science integral to understanding how and why humanity came to use fossil fuels, and how that use came to impact our climate. Before lunch time, we had set each of the songs and rehearsed each a few times. We ate lunch together in the NCAR cafeteria and had the youth sit with top national scientists to discuss their ideas for their solutions for city resilience in the face of climate change. After lunch, two of the CU students took the youth on a short hike into the foothills to give them a chance to be outside. We then brought them back into NCAR to view the display on fossil fuel formation and its impact on climate, featured in the NCAR gallery. Once we reassembled in the conference room, we broke up into four groups to create skits for Act Two. We evenly dispersed adults with youth to work together to identify local solutions to increase their city's resilience. The adults were deliberate in letting youth voices take the lead in this process, simply encouraging and deepening their thinking through asking questions. Groups had at their disposal large square primary-colored pieces of fabric to use as props or costumes in their skits. They were asked to focus on one single solution, frame it locally in the context of Boulder, and to keep their skit to two minutes or less. Once the skits were created, we had time to do several full run-throughs of the entire performance.

At NCAR, youth got to experience expressing themselves in a place of positive social power. Youth expression became a part of the daily operations of this internationally acclaimed center for climate science. The guest of honor for our NCAR afternoon performance was Boulder City's Chief Resiliency Officer with the Rockefeller Foundation 100 Resilient Cities Initiative, Greg Guibert. Each group that created a skit was also charged with writing out their idea on a piece of paper. At the end of the show after the final bow, the youth performers ceremoniously presented an envelope containing their solutions to Guibert.

That evening they performed the show again at University of Colorado for the Conference on Communication and Environment (a biannual event put on by the International Environmental Communication Association), this time on a traditional stage for a group of over 300 academics, journalists, and policy workers who specialize in communication of environmental issues. The response was extremely positive. After attending sessions heavy with policy and challenges, conference attendees expressed a resurgence of hope by watching the performance. When we explained to the students who would be present at the evening performance, the young performers were both nervous and excited. Given the positive audience reaction throughout the show and the standing ovation at the curtain call, the youth seemed to feel their audience's approval and encouragement. From informal discussions that followed, many students came to a deeper understanding of their performance as a form of environmental communication—one that many experts commended as being fresh, hopeful, and highly effective.

Highlight of Youth-Authored Solutions

- Continual blows of alternating axes are chopping down a girl wrapped in green cloth impersonating a tree. A narrator explains that through deforestation we are posing a threat to the stability of our climate, explaining that when a tree breaths, it inhales carbon dioxide and exhales oxygen, which helps to clean the air. Once our mighty tree finally falls, the narrator asks the others what could be done as a solution. One girl suggests replanting more trees and another boy agrees. They all sing "Save the Trees" together multiple times while waving their hands at the fallen tree. She is resuscitated and strikes a triumphant pose as a full-grown, healthy tree.
- Five people enter the stage chanting, "We're going to a meeting, and this is our greeting." They are interrupted by a boy who jumps into the scene announcing he has just arrived using the new and improved public transpiration. A woman interrupts and explains that she couldn't get a babysitter to watch her kids during this meeting for Boulder Resiliency Planning, and wonders if she can bring her kids along. They all heartily reply that she can. Another girl announces she brought snacks. A woman calls the meeting to order, but is interrupted by a young man who suggests they get their creativity cooking with a laughter break. Two people share jokes; everyone laughs, and the meeting resumes with improved focus and clarity. It is decided among them that if more than three unrelated people could live together in one home, it would be way more sustainable (something that was currently difficult to do because of restrictive Boulder City Ordinances).
- People in a neighborhood gathering after a flood (alluding to the Boulder flood of 2013 that was declared a national disaster) attempt to hold a meeting to ensure that everyone has what they need. One elderly woman is offered special assistance with her home. Others agree to form a committee for the long-term response to all the damage. They end by sharing a meal together for sustenance and increased community bonding.

- The actors playing Sol and Foss enter still in costume. Sol announces that Boulder should receive all its energy from her. The audience cheers. Foss glares at the audience with a heavy brow. He apologizes for what he did at the end of Act One but argues that Boulder can't just get rid of him. He reminds the audience all he has done for the people of Boulder, and warns that if only clean-energy use is enforced, only the rich will be able to afford access to energy. Sol agrees to transition to a clean energy future gradually, project by project, first installing solar-powered street lights, requiring building regulations that are Greenstar and eco-friendly, and by starting a youth-led energy education program in their schools.

Lessons Learned and Recommendations:

1. Youth Voices Contribute to City Plan for Resilience

In some ways, the June 12, 2015 Boulder performance most fully realized the ultimate goal of this performance design and experience: for youth authorship through performance to be included in the city's plan for resilience. *Shine* was included in Boulder's official city plan for resilience as part of the Rockefeller Foundation 100 Resilient Cities Initiative. Because Greg Guibert, Boulder's Chief Resiliency Officer, was a guest of honor at the performance at NCAR, he highlighted this performance of *Shine* in the City of Boulder Resilience Strategy. In this official document, the outcomes from *Shine* were used as an exemplary example for Action Item 1.6 titled "Foster Artistic Engagement," which called to: "Engage the creative power of the arts to convey and involve people in complex risk and resilience themes," and goes on to explain how, "Mobilizing action at a broad scale also requires varied ways of communicating complex topics so that they are relatable and actionable to the diverse residents that make up that community."[4] Guibert not only attended the performance but also received an envelope containing youth ideas for solutions. This action of sharing youth-authored solutions likely contributed to the ultimate inclusion of their solutions into the city plan.

2. Validation for Youth Voices

Inviting audience members who have power and influence within the community—such as city officials, journalists, and scientists—to a performance can validate youth expression and can increases the chances that youth ideas, needs, and perspectives will be included in city resilience planning. Rehearsing and performing in places of positive social power, such as a national laboratory (NCAR) or a university (CU), can validate youth voices and their contributions. For many youth in our production this was their first time visiting these spaces. Not only were we surrounded by physical beauty—breathtaking views of the Rocky Mountains at NCAR and a grand ballroom at CU—but the social importance of these places was transferred onto youth expression. The use of these spaces for performance allowed for an inversion of the politics of these spaces. At NCAR, the scientists received information from youth, rather than what usually happens at NCAR: that youth groups visit the gallery to receive the scientific messaging from the displays based on the research of NCAR scientists. At the conference at CU, the usual order was likewise inverted as accomplished environmental communicators were the recipients of youth environmental communication.

3. Live Drumming

Having a live drummer present, in addition to the previously recorded music, contributes significantly to maintaining the dramatic energy during transitions and during non-musical portions of the action. In this case, the composer of the music, Tom Wasinger, was present to provide live accompaniment. His previous knowledge of the show and intimate familiarity with the music allowed him to skillfully accent certain actions, smooth over exits and entrances, and increase the urgency during intense moments. Any adept musician who participates in rehearsals could likely add energy through live musical accompaniment.

4. Community Building Through the Show-in-a-Day Structure

The forming of community through performance was supported by the fact that this production of *Shine* engaged the performers from 8:30 a.m. until 10:00 p.m., which made for a full-day immersion in the exploration of energy and climate as it impacts their city's resilience. It allowed the arc of understanding to evolve over time and with an intensity of focus. Extra-daily concerns were largely pushed aside as we all had surrendered the entire day to the mounting of *Shine*. This race-to-the-finish style of performance also fostered the kind of accomplishment felt by participants in a marathon and engendered a heightened feeling of belonging to this community, and, by extension, our city.

Audience Feedback

I had specifically asked Dr. James White, Professor of Geological Sciences and Director of the Institute of Arctic and Alpine Research at the University of Colorado to note if there was anything in the script that should be adjusted for either scientific accuracy or in the way humanity's relationship to energy was presented. He responded via email: "The show was awesome! The kids brought such energy and enthusiasm to the production, and I loved that they had a hand in telling the solutions part of the story. I thought the telling of the story was even-handed and honest. Kudos for tackling such a difficult topic with such directness."[5]

Cathy Deely, a parent, wrote of the performance: "I love how so much intelligent activism that's going on now is going towards our youth directly, while many adults are still arguing over what's problematic or whether or not certain problems really exist (i.e., global warming). The kids are more than ready to take the baton and run with it. This realization makes me feel much more optimistic."[6]

Casey Middle School and the Sustainability, Energy and Environment Complex at the University of Colorado

Where: Boulder

When: October 2, 2015

Community Partner: Casey Middle School Theatre teacher Julie Metzger, "Inside the Greenhouse," and the Albert A. Bartlett Science Communication Center

Short Description: Casey Middle School youth fully facilitated Act Two by engaging the entire audience in authoring skits dramatizing local solutions. Boulder leaders of science, city planning, grass-roots organizing, and non-profit organizations attended in full-force to celebrate resilience-themed arts within CU's newly completed Sustainability, Energy, and Environment Complex.

Description

The University of Colorado was still putting the finishing touches on its impressive new building, the Sustainability, Energy and Environmental Complex (SEEC), when a cast of Boulder youth christened this new hub for climate science, innovation, and policy with a performance of *Shine*. I, along with CU graduate student Alia Goldfarb, rehearsed with Julie Metzger's seventh-grade theatre class at Casey Middle School from the end of August to the beginning of October twice a week in their Theatre classroom space. At the final half-day weekend rehearsal and during the performance, we were joined by six CU students who performed alongside the middle-school students and helped guide the transitions. Separate rehearsals were held with a CU female graduate student, Ligia Batista, who played the lead role of Sol and fourth-grade male student from University Hill Elementary, MacLeay Bolgatz, who played Foss. Their exaggerated age difference emphasized the notion put forth through the play that fossil fuels are a younger form of energy than solar energy. We performed in the open lobby of SEEC for approximately 150 people. This event was presented by a CU initiative, Inside the Greenhouse, and was co-sponsored by the Albert A. Bartlett Science Communication Center, which is housed in SEEC.

This performance was remarkable in three ways: (1) the performance was featured as part of an entire array of creative offerings that demonstrated the power of the arts for expressing themes of resilience; (2) an extraordinary effort was made to include Boulder leaders of science, city planning, grass-roots organizations, and non-profit organizations involved in resilience planning; and (3) student performers led small groups of audience members in creating Act Two that evening.

1. In the lobby of SEEC was an "Inside the Greenhouse" exhibit of Young Women's Resilient Voices, which included kites and capes painted by young women in Boulder to express their vision of resilience. This was done in partnership with the CU initiative "Growing Up Boulder." Further information on this exhibit and a template for leading a similar event is included on their website http://www.growingupboulder.org/kites-and-capes-of-resilience.html. A 7-minute video on Young Women's Resilient Voices (http://www.growingupboulder.org/resilient-voices.html) was shown while people found their way to their seats that features young women in Boulder sharing their views on resilience. This and the MUSE Studio exhibit were created with young women from Casey Middle School and GeneSister (a Boulder County pregnancy prevention program), with Amistad (an organization promoting empowerment and leadership for Spanish-speaking residents of Boulder). Several of the young women who created kites and capes were also featured in the video and a few were performing in the live performance that night.

Also present in the SEEC space on opening night was the Institute for Social and Environmental Transition-International's (ISET) display of their interactive Prayer Wheels. Two 3-foot tall ceramic prayer wheels were adorned with detailed illustrations, one of the threat of climate change, and one of the hope associated with the resilience of our planet. People were invited to give each wheel a gentle push to watch the stories engraved into the wheels "come to life."[7] Directly following the performance of *Shine*, Bateria Alegria, a local Brazilian Samba band, shook the rafters with their drums as the audience enjoyed conversation, delicious food, and dancing. There was an affirmation of the power of the arts for invigorating our community and to specifically express what resilience means to us, why it matters, and how we can creatively plan for it as a community. Also, this expression outlasted the night-long performance as the kites and capes exhibit remained displayed in the SEEC lobby for nine months following the performance.

2. Months before this event, Boulder community leaders were contacted and asked to serve as expert guides in creating solutions for resilience planning at the event during the creating of the Act Two skits. They were told that they would be paired with a youth to help a community group identify a local resilience issue and create one viable solution to perform for the audience. Their specific job was to reflect back to the group if the idea seemed plausible, not to lead the group since that was being done by the youth. We received commitments from ten local experts, including: Nobel-prize winning Physicist Eric Cornell; Director of the Institute of Artic and Alpine Research at CU James White; grass-roots organizer with the Climate Culture Collaboration (C3) Joellen Raderstorf; Senior Environmental Planner for City of Boulder Brett Ken Cairn; and founder/director of ISET Marcus Moench.

3. The Casey Middle School theatre class extensively rehearsed how to facilitate and lead the audience in co-creating Act Two of *Shine*. After the final tableau of Act One, ten Casey students spread out evenly around the lobby, each holding high a poster with a number on it ranging from one to ten. Audience members were asked to gather into ten equally sized groups with one of these students. Each student was assigned an expert guide, who was an adult community leader and stood with the student holding the number. We explained to the audience that they were going to participate in identifying a local resilience challenge, devise one solution, and create a skit to communicate that solution.

Although everyone in the group was asked to contribute to the process, not everyone had to perform. Before gathering into groups, we provided an example skit performed by two CU students and three Casey students that identified the carbon impact of family vacations as a local challenge and stay-cations as one possible solution. They enacted two parents talking with their kids to plan a stay-cation in Boulder, which even allowed them enough financial savings to buy new bikes for the kids. The groups were given ten minutes to create their own skits that were to be approximately one to two minutes in length.

The transitions from one skit to another were spirited and timely, covering space in a fun and efficient manner. These transitions were truly youth led, as the youth had rehearsed using two refrains of the "Bounce forward, rebound, that's my resilient town" chant to switch from one group to the next in the designated performance space. I marveled at how this was, beyond being entertaining and fun, an extremely efficient way to get input and involvement from 150 people in such a short amount of time. (a) They organized into small groups to consider a local solution to energy-use impacting our climate and how to move our city towards being more resilience. (b) They decided how to express that in a skit that contextualized the solution in our city. (c) They presented this to the larger group through a short skit. (d) Finally, they were able to witness and celebrate each other's ideas. It was a great way to involve all ages in the action, as little kids in the audience got up and performed alongside their parents. Once the skits were all done, the cast distributed colorful paper streamers for the audience to wave during the final number, *Shine*, with the hope that the audience might be more likely to dance if they had a streamer to wave to the music.

Description of Youth and Community Authored Solutions:

- A mom and a dad are looking on at their three children who are all playing video games on their phones. The parents discuss their regret that their children spend so much time indoors playing games instead of outside in the beautiful mountains. Finally fed up, the mother takes the phones away from the kids and announces they are going to play soccer, which they do, and all have fun.
- A narrator describes that these old buildings—represented by part of the group in pairs making a roofline with their arms—are bad for

the environment and use a lot of energy to heat. A pedestrian near one of them coughs because of all the toxins in them. The narrator then shifts our attention over to these new buildings—represented by others in the group—that are solar powered. These are much better for the environment and will make our world a better place.

- Two rich people are in a very big house bemoaning all there is to do for such a big house, how it is too much for just them, and how it is big and lonely. Two people run across the stage waving their arms like torrential waves representing the Boulder 2013 flood coming through town. Many displaced people follow in their wake. The rich people invite the displaced people to come live with them forever and ask each what they will contribute. Suggestions include Thursday night pasta dinners, gardening, and cleaning windows.
- A woman asks, "Who wants to go to the SEEC opening party?" to which everyone responds enthusiastically, "yes." She announces she will be taking the bus and invites everyone else to join her. Two people decide to drive their cars instead. The people on the bus chant "fun on the bus, fun on the bus" and make their way to the event. The two drivers buzz around in their cars until they crash into each other and have to abandon their vehicles. The bus approaches and offers them a ride, and they get on and add to the chant, "fun on the bus."
- Five people lie on the ground in a row representing excessive cement coverage of the city. A narrator asks if anyone will help dig up this concrete to make way for planting trees. Once cleared, some people who represent trees crouch on the ground. The narrator announces the rain and the trees grow, one man even lifting a person in the group up onto his shoulders with that person's arms branching out up to the sky (Fig. 3.4).
- Three people stand along the back of the stage with their arms up to form a "V" over their heads. Another woman starts hammering something onto one of the rooftops announcing that she is installing solar panels. When another woman comments how expensive solar panels are, she responds that Boulder has a program that makes it very affordable to go solar. She then asks the other roof if she would like solar panels too, to which she nods yes enthusiastically.
- One young man announces he is a farmer. The others in his group line up to gain entry into the farmer's compost pile. He asks each

Fig. 3.4 Skit dramatizing the growing of a tree. Photo by Rebekah Anderson

one what it is. The grass clippings are allowed, as is the old rotting banana peel, and the watermelon rind. The chicken bone is asked to go to the county compost instead, but the broccoli is allowed in. The compost pile is shoveled together as the group forms a clump. They decomposed down to the ground together. From their center a young girl hiding within pops out and announces that she is a flower.

Lessons Learned and Recommendations:

1. Get Community Leaders to Attend

One of the primary goals of this project is to activate youth in contributing to their city's plan for resilience based on their needs, perspectives, and ideas. Another goal is for this youth contribution to be received, considered, and included in the city's plan for resilience. For this to happen, it helps to have key city officials, scientific leaders, and community organizers in attendance. For this SEEC event, our success in getting the commitment of so many busy and prestigious people was attributable to identifying people who were personally and/or professionally invested in the goals for the event and asking early. It pays to ask months before the event and with enthusiasm, specifically articulating why their presence is so important. I found that by telling our guests that their presence would greatly contribute to the empowerment of youth, it was pretty hard to say no. In addition, giving them an actual function in the event—such as being an expert guide—seemed to guard against them cancelling as their busy schedules threatened to overtake them as the date of the event approached. We also sent out a reminder email a month and a week preceding the event. It also took a lot of pre-planning to ensure attendance by city officials, parents, and the wider community. I met with ISET a month before the event to learn more about their work and invite them to share their art. It took a fair amount of explaining of the event before Eric Cornell felt comfortable committing to being an expert guide, perhaps because he is an acclaimed scientist weary of being used as a figure to advance a specific political agenda. Once he did participate, he had a lot of fun and felt positive about his participation. Ensuring the attendance of leaders within the community goes a long way in validating the experience for youth and the community at large. One of the goals of this performance experience is to bolster the community opinion of youth contributions to a city's resilience.

2. Support the Creation of High-Quality Skits

By high quality, I mean skits created through a creative process that benefits from a multitude of perspectives and that produces viable solutions that address a specific local resilience issue or issues. Even if the solutions are outrageous, they benefit from being informed by what is possible in scientific terms and by being improved by scrutiny and critical reflection from a group of people with different life experiences. Given that criteria, this set up created exemplary skits. Performing an example of an effective skit for the audience provided a model for them to use in terms of tone, length, freedom, and focus. The pairing of youth with a designated scientific expert contributed to skits putting forth realistic solutions. It also helped to work with the youth leaders in rehearsal on facilitation skills for leading groups in creating skits, such as: (1) lead through questions to engage your group; (2) add more enthusiasm to the process; (3) keep your group focused on identifying a solution and then create a skit instead of just debating issues; and (4) encourage contributions from everyone in the group. Ultimately a high-quality skit also inspires both the creating group and the audience to want to take action on the solution offered.

3. Youth Can Effectively Facilitate Community Engagement in a Lively and Efficient Manner with Planning and Rehearsal

I had no idea that the youth would be so capable in herding the audience into their groups, focusing their group's attention on their task of creating a viable local solution, getting them to rehearse a scene to communicate that solution, and leading them to the stage and back, quickly and in style. In the weeks preceding the performance, every time we did a full run-through of Act One, we also rehearsed the student facilitation in Act Two. That seemed to really pay off during the performance when they so quickly and efficiently facilitated the community involvement. This aspect may have been one of the most impressive outcomes of this event and certainly made an impression on the invited guests. Many people commented (especially the expert guides who had been specifically invited to help the groups along by being part of a group) that the creation of the skits was truly youth led.

4. Youth-Focused Events Give Adults Permission to Play Again

Youth seem uniquely situated to lead their community in embodied, participatory creative exploration of local solutions to climate and energy related challenges. Adult audience members seemed to be reminded of a time when they played and tried new things. The fact that it is youth led seems to give people permission to play. I also sensed that the audience was grateful for the effort the youth performers had put into rehearsing and performing Act One and were inspired to match that level of energy and commitment in authoring Act Two. Truly the skits put on by the groups were outrageously imaginative and physically committed.

5. Listen to Youth Participants

As described above, young women in GeneSister who are primarily Latina and generally under-resourced in economic terms, participated in an art workshop to generate rising phoenix capes to represent resilience. The hope was that they would feel validated to have their artwork displayed in SEEC, a place of positive social power. My community partner in this project who directs GeneSister, Maya Sol Dansie, even offered transportation to the young women to and from the event since she has access to a 12-passanger van through Boulder County. The NCAR scientist Paty Romero Lankao had translated the script into Spanish and a CU graduate student Goldfarb, had created slides of subtitles that we were planning to project on the back wall during the performance. As the event approached, Maya informed us that, though she had tried mightily to rally her girls, none of them were going to make it to the event. Although neither she nor I knew the exact reason why no one wanted to attend, we suspect that it was the perceived foreign, intimidating, and maybe boring perception of this space that was part of the detraction. Perhaps we needed to incentivize attendance more thoughtfully in terms of what was actually valued by the young women and not so much on what we thought would be of value for them. Including them in a discussion on that more directly and earlier on in the process may have increased their desire to attend and be publically acknowledged for their artwork.

Feedback from Performers and Audience Members

The following feedback is from interviews with three Casey Middle School students who acted in the performance. Leela Stoede in seventh grade said, "You can learn a lot from action. Learning from other people and hearing their ideas got me even more into it. It opened my eyes to the fact that the people who did the least amount of damage are the most hurt."[8] Charlotte Gerrity in eighth grade said, "The arts are a way for me to express myself and how I feel, because sometimes writing a paper just doesn't connect with me as much as performing it. I tried to make it fun for the audience, but also to challenge the audience that this isn't just a play, but this is actually happening. When I was leading my group in the skit, I tried to get everyone involved so I asked for volunteers. They were really into it. I feel like people didn't realize that we young people have a say, and I don't think people really took us seriously until we showed them that we could lead this whole thing."[9] Lucinda Stewart in seventh grade: "Being in the play felt like I could express how I think our city could change, that it should just stand up to whatever problems it might face. I feel like the arts are the most powerful means of communicating because you can get your message across in a way that can get to every kind of person. They can see the problem and they can see what it is doing. Dance is a form of giving your message. When you dance, some people might see one message and other people might see another, and you'll get to lots of different people that way. I felt like the play was very detailed in how it portrayed how climate change was caused and how it is now our responsibility to fix it. I felt like at the final scene where the big bad people broke the fabric of community that we caused this. We did this. It wasn't anyone else. We did this, and it's our responsibility to fix it."[10]

Brett KenCairn, Senior Environmental Planner with the City of Boulder, wrote the following after attending the performance and serving as an expert guide during Act Two. "As a climate policy professional, I spend most of my time working with information that is disturbing at best and terrifying at worst. A significant part of the challenge we face in working with the public in developing effective responses to our climate crisis is the increasing emotional and psychological fatigue that most people feel about what appear to be rapidly diminishing prospects for creating a liveable, prosperous, and equitable future. The *Shine* experience is a powerful tonic and antidote for this fear and despair. That it engages our youth as part of this process is part of its beauty, and part of its wisdom. To be reminded by these young faces both why we are doing this work and that there is every reason to have hope that when we come together around our deepest and most cherished values—love of place, family, community, and the natural world—we have what we need to transform our situation. Thank you for the beautiful and nourishing work. I am so glad you will be able to share it with other places."[11]

Phaedra C. Pezzullo, Associate Professor of Communication at CU wrote the following note after attending the performance with her pre-school-aged son, "Thank you for creating such a participatory and generative space for resilience tonight. Insightful, fun, and my child was joyous."[12]

Shelly Sommer, Information and Outreach Director of Institute of Arctic and Alpine Research at CU wrote that *Shine*, "was an uplifting performance that fully engaged our diverse audience of students, scientists, local families, and kids. Perhaps more remarkably, it also helped that audience feel safe enough, and inspired enough, that they enthusiastically participated afterwards in short discussions and skits about how to make our town more resilient. I have never seen a better instance of a performance that actually inspired the audience to take action."[13]

New York City, New York

Where: New York City
When: October 23, 2015
Community Partner: Hunter Elementary School, The Urban Thinkers Campus, The New School (other *New York Convening on the City We Need* conference participants included the Sherwood Institute, Huairou Commission, Columbia University, University of Pennsylvania, Ford Foundation, Lincoln Institute of Land Policy, International Accountability Project, United Nations Major Group for Children and Youth, the Nature of Cities, and J Max Bond Center On Design For Just City)
Short Description: With Arthur Fredric's choreography and direction, New York City youth performed *Shine* as the culmination for a major conference building up to the United Nations conference for urban development, "Habitat III (Fig. 3.5)."

Description

"I'm so relieved this wasn't a finky kid thing," said conference-organizer Crista Carter, exhaling after the performance of *Shine* at the Urban Thinkers Campus: *New York Convening on the City We Need*. Carter had taken a leap of faith by scheduling *Shine* for the final plenary session of this conference—an initiative for the United Nations Habitat. Conference attendees, exhausted from days of serious deliberations, appeared to have their hearts lifted back off the ground by the energy emanating from the youth performers. During the final number, the audience rose to its feet, everyone smiling, and the entire room infused with rejuvenating hope. Many people commented that this performance was a great culmination to very heady, intellectual days of deliberations.

Directing and choreographing this production was Arthur Fredric, a former Broadway performer who worked extensively with Jerome Robins and currently teaches for the National Dance Institute (NDI) in NYC. Fredric is a master at leading youth in schools—from every borough and neighborhood—to achieve their highest level of artistic excellence through dance. Fredric, along with his partner in life and dance, Lisa Denning, had already previously visited CU as guest artists to advise on the creation of *Shine* in its earliest incarnations. They had helped build some of the basic movement phrases we used in the Navajo Nation and

Fig. 3.5 Fossil fuel flags Photo by © 2016 Steven Sutton, DUOMO

in Boulder. This New York performance marked the creation of original choreography and direction by Fredric for *Shine*, which significantly increased the artistic caliber of the production, and made its inclusion in this international conference appropriately well received.

Fredric had worked with students at Hunter Elementary for years through NDI, so he was able to handpick fifth and sixth graders whom he thought would benefit from the experience and contribute to the artistic excellence. In rehearsal, he worked the youth performers with rigor, interrupting himself occasionally for comic relief. They both met the challenge and groaned at his jokes. In the lead role of Sol was Chelsea Hackett, a Ph.D. student at New York University, Foss was played by a high school student at Hunter named Emmet Smith, and

Topaz Hooper—a recent CU alumnus who had travelled to the Navajo Nation when the show was first performed and at NCAR in Boulder—reprised her role as Seed Sower. Composer for *Shine*, Tom Wasinger, supplied live accompaniment with another NDI musician Tim Harrison.

In the week preceding the performance, Fredric held two 2-hour long rehearsals at Hunter Elementary. During this time, he taught them the basic steps—some of which were familiar to them through past projects. We then met on the Saturday of the conference at New York University to rehearse from 9.00 a.m. to 12.30 p.m., after which we walked over to The New School and did our run through in the Starr Foundation Hall. An expert on climate, Joshua Sperling with the National Renewable Energy Laboratory, was attending the conference and took time out during our rehearsal at the New School to talk with the Hunter student performers about the science in the show, and he performed as a stagehand in the performance at the conference.

In fact, it was Sperling who introduced me to Carter, who is with Urban Policy Analysis & Management at the New School and was one of the primary co-ordinators of this event. Sperling envisioned a performance of *Shine* at this event as a way to include the contributions, presence, and perspectives of youth in these important deliberations. This conference by Urban Thinkers Campus, an initiative of UN Habitat, was conceived as an open space for critical exchange and as a platform to build consensus between partners engaged in addressing urbanization challenges and proposing solutions to urban futures (www.worldurbancampaign.org/urban-thinkers-campus-city-we-need). Many similar conferences were held all around the world to gather momentum for Habitat III, which occurred in The Republic of Ecuador in October of 2016, just one month after this performance in New York. The United Nations organized Habitat III, the third in a series that began in 1976, to "reinvigorate the global political commitment to the sustainable development of towns, cities and other human settlements, both rural and urban."[14] The room in The New School in which we performed had blond wood on the floor, walls, and ceilings. Lines of chairs were pushed together to seat over 100 attendees in formal attire. When the opening drums marked the beginning of the show and the performers entered from the back of the room, dancing in as ancient plants and animals, the space was changed. Throughout the show something palpable was integrated into our shared humanity. The future projections for the city were being performed by the future holders of that city. It wasn't just the hope for the future they represented, but the current-day critical engagement and action they were demonstrating with the investment of their bodies, imaginations, and voices. Youth performers opened a door to the joy inherent in moving from concern to action on your most passionate concerns.

Description of Youth-Authored Solutions

Acceptance into the Hunter Elementary is highly competitive, due to its academic excellence, so the students we had to work with tended to be extremely bright, motivated, engaged, and creative. This was reflected in the solutions they created. I utilized a time-efficient technique for them to create their skits in a manner that would engage each student in the process of identifying problems, critical reflection, and problem solving. The performers gathered into three groups of four people. I asked everyone to take one minute to think of a local solution to a local resilience challenge. Each person was then asked to share their idea briefly with their group. Groups were asked to consider all of the ideas and, together, come to consensus on what solution seemed most viable, achievable, and effective. This process supported them in working together to come to consensus in putting forth a unified solution.

- The spokesperson for the first group explained how they proposed getting rid of all the Metropolitan Transit Authority buses and replacing them with trains that go above ground to not waste so much gas. Two others in his group made train tracks with their arms and a girl stood between their arms turning the wheels of the train with her arms.
- The second group's spokesperson proposed having airplanes powered by solar panels, homes fueled by wind power, and models of cars that could be made electric. During his description, his fellow group members flew like airplanes, joined their arms to represent solar panels on a rooftop, and mimed driving a car respectively.
- The spokesperson for the third group explained that if people emit too much polluting gas, they will have to pay. If people use more than what is allotted, they get charged $50, and if they are using way too much they get charged $100. That might make more people switch to solar energy.

Lessons Learned and Recommendations
Increased Artistic Excellence Contributed to Acceptance/Reception of Youth-Authored Contributions in High-Level Gatherings

Having Fredric also direct the show with Denning contributed to its artistic cohesiveness and excellence overall. One purpose of this New York performance was to ensure that youth voices were included in the build-up to Habitat III. It is likely that nearly every person who witnessed the Hunter youth performance, carried some piece of their spirit into the work they do and to the final Habitat III conference, if they attended. The connection between urban development and youth expression was made real for attendees by the inclusion of youth performance at the *New York Convening on the City We Need* conference. Increasing the artistic excellence of the production—through Frederic's choreography and direction—likely played a big part in the positive reception of *Shine* at this event. By association, the artistic excellence likely increased the extent to which youth solutions were received and seriously considered. Due to the nature of embodied performance, the audience not only received the chance to see the issues through the eyes of youth, but also was given a visceral demonstration by youth of engaging their entire selves in responding to the challenges of our urban futures.

There was an entire day, October 15, 2016, at Habitat III in Ecuador that was dedicated to youth entitled "The Children and Youth Assembly." This took place to further include the voices of youth in urban development. This one-day inter-generational and inter-stakeholder forum provided a platform for young people to share experiences and knowledge, showcase solutions and initiatives, and develop partnerships to make cities and human settlements more equitable, inclusive, safe, resilient, and sustainable.[15]

Feedback from Audience Members

Morana M. Stipisic, who teaches Urban Design at Columbia University, wrote, "I must admit to have been rarely amazed by what I saw—and in this city, it is not easily achieved. Not only that the project itself is so meaningful, but the way those young students happily engaged into performing and accepted the material in a mere day serves as a proof of the right track of thinking. One thing that I deeply believe in is the fact that the children are the ones who stand a true chance of finding solutions for the betterment of our urban future. As Albert

Einstein said: 'We cannot solve our problems with the same thinking we used when we created them'."[16]

LONDON, GREAT BRITAIN

Where: London
When: January 12–22, 2016
Community Partner: University of East London Department of Drama, Applied Theatre and Performance, and Riverside School in East London
Short Description: A week-long residency with theatre students at the University of East London culminated in a performance of *Shine*. This prepared them to co-facilitate a residency in an East London school with their entire Year 7 (sixth-grade) class for two whole-school assembly performances. In the weeks preceding the performance, several Year 7 teachers collaborated for a combined, multi-disciplinary investigation of a single subject through the arts centered around *Shine* (Fig. 3.6).

Fig. 3.6 Foss and Sol in a bout of sibling rivalry. Photo by Melisande Osnes

Description

Since the theatre departments at University of Colorado (CU) and the University of East London (UEL) have a student exchange program, as a CU professor I approached UEL with the idea of hosting a *Shine* residency. Dr. Ananda Breed with UEL welcomed the idea. In addition, she introduced me to the Head of the Music Department at Riverside School in East London, Soren Ramsing. He had previously attended a UEL open house to find willing collaborators for integrating applied theatre into his school to facilitate a combined, multi-disciplinary investigation of a single subject through the arts. Breed made the connection between Ramsing, myself, and *Shine*. Once Ramsing and I spoke, possibilities sparked, and the plan between UEL, Riverside, and myself expanded and took shape over the course of several months of planning.

University of East London

During my first week in London I conducted a week-long intensive workshop on *Shine* for 19 first-year UEL performance undergraduate students with the objectives of: (1) facilitating UEL students in performing *Shine*; and (2) through their experience, to prepare them to facilitate the Year 7 students of Riverside School in performing *Shine*. Also participating in this UEL week-long workshop were two UEL third-year students who were both former CU exchange students; Johnny Whiting and Patricia Akoli—who portrayed the roles of Foss and Sol respectively. While at CU, Akoli travelled to the Navajo Nation with our mounting of *Shine* at Tuba City High School, so she was familiar with the show and its aims. I met for four hours a day for four days with the UEL students in a performance studio at UEL from January 12–15, 2016. For each session, we began by warming up together using many of the exercises and activities included in Chap. 2 of this book followed by some activities specific to the understanding of resilient communities.

Embodying a Resilient-Community Activity

One of these resilience activities, Resilience in Motion described at the end of Chap. 2, begins by asking everyone to stand in a long straight line. A person at one of the ends is designated as the lead person and turns to face the rest of the people in the line. The exercise continues until everyone is grabbing hands, pulling, grabbing, pulling, so the line becomes like one big organism moving together. In London, for an added challenge, we did this same movement pattern in a circle (this arrangement only works with an even number of people participating.)

After doing this activity in a line and a circle, we shared the following questions to reflect on the activity:

- What does it take for individuals to work together as a working community?
- What can make it fall apart?
- How do different designs for human communities, the line versus the circle, either promote or change the capacity of a community to work together?
- Why did the attempts sometimes fail?
- How did it recover?
- What can help a community or individuals within that community recover or bounce back?
- When doing this activity, how did we resemble a resilient city?

After the opening activity, we launched directly into rehearsing the play using the curriculum in Chap. 2. Throughout, we critically reflected on our process and discussed how we might draw from our experience of mounting the show to inform how we planned to facilitate that same process for the students at Riverside School the following week. The material of the show easily held the interest of the university-level students. Slithering across the room as ancient animals and dancing the various parts of photosynthesis as ancient plants maintained their spirited commitment.

Reflection

At the end of each day we reflected on the theme of resilience using a variety of activities such as the following. The group laid on their backs with their heads all towards the center of the circle and closed their eyes. The facilitator guided the group in silently considering each prompt, reading each of the following with a period of silence between each:

I feel resilient in my own life when ...

I contribute to the resilience of others by ...

In order to feel resilient, I need these things from my family, friends and/or community...

My resilience will contribute to my community by ...

I rely on my city to be resilient because...

On the day that we created the solution skits for Act Two, we went around the circle, each sharing a resilience challenge specific to London, including both climate shocks and social stresses. Challenges identified included: flooding, lack of affordable housing, pollution, sea-level rising, and terrorism. In groups of four, they chose an issue and a solution for the content of their skit. To scaffold the creation of the skits in a manner that emphasized physicality above just the verbal communication, I used the following activity to specifically support embodied expression. This following exercise is based on Image Theatre as developed by Augusto Boal that is written about in his seminal book, *Theatre of the Oppressed.*[17]

Image Theatre as an Exercise for Devising Solution Skits

Once each group had decided on an issue for their skit, I asked them to create three distinct images: (1) an image of the problem, (2) an image of the solution to the problem, and (3) the transitional image, or an image of the action that got them from the problem to the solution. By "image" I mean a frozen scene made up of their bodies that physically communicates each prompt. I instructed them to portray a specific manifestation of the problem. For example, if the problem is homelessness, the image of the problem could be a man sleeping in a park with police attemtping to arrest him, the image of the solution to the problem could be the man working in a job helping to clean up the liter in his city with others, and the transition could be a city hall meeting in which police and citizens are considering transitional employment for formerly homeless people.

Once each group created their three images, they took turns sharing these with the others, one group at a time. First, they showed the image of the problem, then the solution, and finally the transition. They were not allowed to use any words when presenting these nor were they allowed to announce what their issue was. Once each group was done, I asked the others to reflect to the group what they saw. This gave each group a chance to hear what was communicated clearly and what might need more description or clarification. I then gave them time to create a skit that was about one to two minutes long. Urging them to be playful with the creation of their skits, I told them not to over-think them, but, rather, to get on their feet and actively work through the creative process. To support this, I only gave them ten minutes to create their skits as a group, after which time they each shared their skit, and we all shared feedback for improvement. I began with the Image Theatre exercise to emphasize the embodied aspect of their communication.

I had brought along five huge colored balloons—each one 36 inches in diameter—and gave one to each group to inflate and then decorate with a visual representation of their solution using wide-tipped, water-based paint markers. The fifth balloon we decorated with the logo for the 100 Resilient Cities Initiative. While each group performed, someone from another group held the balloon up behind the enactment of their scene to associate this balloon with their solution. At the end of the show, directly after the final song *Shine*, we launched the balloons into the audience. This was intended to symbolically invite the audience to participate in keeping these ideas afloat and to represent the students launching their solutions out into the world. It also provided a feeling of celebration and visual spectacle and got the audience out of their seats and actively participating. On the evening of January 15, when the UEL students did an informal presentation of their version of *Shine* for an invited audience of other students, professors, family, and community members, the balloons were a delightful conclusion to the performance. While performers and audience converged to talk amongst themselves after the show, they batted the huge balloons back upwards when they happened to float down near them, while the few kids in the audience diligently awaited the descent of the balloons and ran to push them back into flight.

Riverside School

In the months preceding the performance of *Shine* at Riverside School, Ramsing, as the Head of the Music Department, had been at work enlisting other teachers to participate in this week of focus on resilience that would culminate in the Friday performance of *Shine* by the entire Year 7 class. Soren referred to this effort as a "cross-curricular project week." He hoped this project would serve as a model for other cross-curricular project weeks that could likewise find their culminating expression through the arts.

The five participating Year 7 disciplines were:

- Music: Ramsing took several weeks of class to teach the students the songs for *Shine*. He also engaged the students in discussions of how music can serve as a tool for conveying the feelings and emotions associated with scientific, civic, and cultural advancements.
- PSHE: The PSHE class studied the concept of resilience and their city's participation in the Rockefeller 100 Resilient Cities Initiative.

Each student wrote a letter to their Member of Parliament for Barking about resilience, expressing their best ideas for how to make London a more resilient city.

- History: Students explored the development of energy use that led to the Industrial Revolution. They explored the many social changes that accompanied this major shift in production and labor.
- Science: The science of climate change was explored with an emphasis on the greenhouse gases that trap heat in the atmosphere and cause the temperature on the planet to rise. They also explored how the greatest human-produced source of greenhouse gases is the burning of fossil fuels for electricity, heat, and transportation.
- Art: The art teacher engaged all of her classes in creating the costumes and properties needed for the performance including a papier-mâché dinosaur head and tail, ancient plant and animal capes made of a tan gauze fabric with colored pictures of ancient plants and animals stapled onto them, stalks of rolled newspaper for the *Harvest* dance, seeds made of cut pieces of tissue paper, strips of paper decorated for the *Weaving* song, flags representing many ways fossil fuels are used in London, carbon made of cut pieces of black tissue paper, and a fireplace and logs sculpted from paper.

The Monday morning directly following the UEL residency, I met with the Year 7 class in their school to introduce myself and teach them some of the movement that would accompany the many songs Ramsing had been teaching them. On Friday of that week, the 19 UEL students joined me at Riverside School for an all-day rehearsal with 135 students to perform the show for two school assemblies, that reached over 525 people, mostly other students, but also teachers, administrators, and some parents. Riverside School is in a working-class neighborhood—with a majority of first, second, or third-generation immigrant families, many Muslim, and many from South Asia or various countries from the African continent. One of the four classes was described to us as including the kids with behavioral challenges and learning disabilities. Five UEL students were assigned to that class to be sure to meet all the needs of the students. Their teacher was also present the entire time to oversee the students and contribute towards the project. We had previously decided they would portray the ancient plants and animals and the dinosaur. For the performance, dressed in their gauze capes, the UEL students who partnered with them wore the green suits and blended in with them helping to guide them through their movements during *Long Time Comin'*, the *Harvest* song, and portraying the dinosaur. The rest of the three classes also had UEL students equally distributed to guide them in their respective portions of the show. Each group of UEL students was charged with the challenge of teaching their class the movement for their portion of the show and in guiding them in creating a skit proposing their best solution for a resilient London. For each group, the UEL student performed alongside the Riverside students to guide them.

Each group of UEL students had a sheet outlining what warm-up exercises they could do with their class, a script, what part of the play they should be rehearsing with their Riverside students, and a reminder to engage the students in creating their own 2-minute skit. Beyond that, the UEL students figured out how to execute all of this in their own way based on the discussions and techniques we had practiced the previous week. The UEL students worked with their student groups in their classroom for the first 3 hours of the day, after which we all assembled in the gymnasium to rehearse the spacing and logistics of the performance together, to familiarize the students with the parts by Foss and Sol, and to rehearse the final number *Shine*. To gather their attention, Ramsing clapped a rhythm and the students would all clap back, using the same rhythm. To organize such a large number of students, we had them sit in groups along the two sides and the back of the stage area. That way they easily rose with their guiding UEL students when it was their turn to perform, and just as quickly returned to their spot.

Due to the challenge of organizing all 135 students for a performance in one day, we inflated the balloons but didn't get the chance for the performance groups to decorate each with their solution. Instead, we launched the balloons mid-way through the final song of the last performance, and near mayhem ensued. Ramsing and I had underestimated the irresistible attraction children feel towards huge balloons. Throngs of Riverside performers rushed to hit the balloon back into the air, nearly knocking each other to the ground. The spectacle and appeal of the balloons made for an exciting, but a bit dangerous, ending. Luckily no one was hurt, though the UEL students did gather the balloons back soon after the conclusion of the song.

The overall performance was outstanding—if a little rough around the edges. I was impressed that each of the two audiences of over 200 students each, all sitting close together, cross-legged on the floor, respectfully gave their attention to these brave performers all through Act One and even through Act Two, which was often difficult to hear and suffered from performers unknowingly putting their backs to the audience from time to time, while performing their skits. The audience seemed to follow the story of Act One based on their eyes staying on the action, their respectful lack of talking amongst themselves, and their responses of laughter at appropriate moments. They likewise appeared to be authentically attentive during the skits in Act Two, perhaps even a bit more so than during the first act, evident by their bodies leaning forward slightly, their facial expressions of interest and delight, and their generous applause at the completion of each skit. This audience response also reflects highly on the school faculty and staff for teaching positive audience manners.

Description of Youth-Authored Solutions: Solutions Put Forth by Students at UEL

- Three people complain about traffic in the city, "It took me ages just to get to work." They banter on about the nuisance of driving, then the main guy shows the others a website on his phone that calculates a safe bike path from your home to your destination, allowing you to travel through parks and on protected bike lanes. The first woman sees him do this for her, she responds positively stating she has a bike in the shed but that she was previously too afraid to ride it on the streets. The other woman too agrees this is a good idea. The first woman says to the others that perhaps they could all try this out this weekend and by doing it get to know each other better. They all agree, turn together, and face front miming riding their bikes side by side, enjoying the ride.
- A woman at a grill yells out "get your burgers here," and another orders a double beef burger with bacon. A bystander says; "stop," and they freeze. She comes forward and asks the audience if they are aware of the impact of animal agriculture? Do they know that it takes 65 gallons of water to produce a single burger? Can you imagine how many showers you would have to take to equal 65 gallons? And that if we all became vegan we could reduce greenhouse gas emissions by 15%? Then she walked back to the scene, snapped her fingers while miming eating a sandwich. The guy who had ordered the double burger asks her what she is eating, and she says a veggie burger. He says, "Right then, I'll have one of those."

- Three women announce that they will go to a local school each week to use interactive performance to explore social issues identified by students as important. They then pretend to be announcing their session at a school gathering. "Remember us from last week when we came and led you all through exploring social issues through theatre? Well, we're going to do that again this week." The first student walks up to a group of students and says, "Last week you said you wanted to talk about mental health issues so let's start with a group discussion," and she freezes as she commences with that. The second student announces to her group, "and you said you wanted to discuss issues around racism," and they all freeze. The third student says to her group, "And we'll discuss issues of bullying around LGBTQ."
- One man stands in the center looking like the balance of justice. The woman on the left declares her thirst, and she is immediately offered a glass of water, which she criticizes as too warm and discards. The other lowlier woman on the right asks for water and is hushed. Again, the first woman asks for water and is immediately offered another glass of water, this time to her liking. The other, lowlier woman again asks for water and is hushed with greater irritation. Then the center male announces as a newscaster that floods have impacted London's ability to get fresh water to some neighborhoods. The lowly woman begins bailing the water that is gathering in her home, the other woman notices this, pauses, considers, then offers her clean fresh water to the other woman and begins helping her scoop the water out from her home. The center man watches this exchange, kneels down, and also offers to assist. The first woman says, "Thanks, we can do this if we all work together."

Solutions Put Forth by Riverside School Students

- One girl stands in the middle of the stage. Other students drive around her beeping their horns noisily as though in a turnstile. She begins to cough and finally protests, "Too much smog and too much pollution; stop!" Once they do, she leans over to one car, fiddles with its mechanical parts, and announces, "There, I've transformed you into an electric car so you will make less pollution."
- Some people are standing around. Some students drive by as though they are cars. The people standing there begin to cough

and one of them falls to the ground. Another reprimands the cars, telling them they have caused this person to have an asthmatic attack. He suggests they all use Segways instead since they are much cleaner and so cool. All the drivers get out of their cars and begin to move around the stage as though on Segways, having much more fun.

- A girl lies on her side on the floor, center stage with her arms close to her body. A boy listening to music on his headphones walks by and tosses his trash saying "I'll just throw this into the river." Another girl passes by and also litters in the river. A group of concerned citizens enter and hold hands around the river, saying how she looks sick and dirty. Two girls hold their arms out in front of them in a circle representing a trash bin. The others begin to pick up garbage out of the river and deposit it in the bins. The boy listening to music returns and is about to toss his garbage into the river again, but this time is stopped. Someone explained that the river is quite dirty and that they are cleaning it. He refrains from littering but does not stay to help. The others continue on with their work.

Lessons Learned and Recommendations
1. The Power of Seeing Ourselves Reflected

Even in such a challenging setting, with cramped and uncomfortable seating, and mediocre sound, what successfully held the students' attention seemed to be watching youth from their own school talk about issues relevant to their city. If the Riverside students aspire to attend university, many of them would likely attend UEL. Most of the UEL students facilitating this project at Riverside were from the same general area as Riverside School, thus making the visiting UEL students both role models and maybe even, local heroes. Given that nearly every Riverside student takes art class, most audience members were able to witness their own creative work get animated through performance. Certain students would have seen the papier-mâché dinosaur head they helped construct, thump across the stage as a multi-person Tyrannosaurus Rex. Others would have noticed the tissue paper they cut into pieces get strewn across the stage as black carbon. Add to this references to local rivers, area parks, and common social problems in East London, which all seemed to contribute to the holding of the students' attention throughout the performance.

2. Combining a University Residency with a Secondary School Project

Combining a university residency with a secondary school project was highly effective. Like both performances in Boulder and the performance in Tuba City, this mixing of multiple ages of youth not only allowed more to be achieved in a smaller amount of time, but it had multiple other benefits. The university students gained experience facilitating youth in arts-based community engagement as well as experience assisting in mounting a show with the youth. It benefited the school because we had a much higher adult-facilitator to student ratio than we would have had if Ramsing or I were facilitating the performance alone. The older university students served as role models for the younger students through their spirited concern for their city's resilience and their commitment to the performing arts.

Feedback

Soren Ramsing, the host to this project at Riverside School and Head of the Music Department, wrote after the performance, "Thank you for helping our school lay down the tracks for future cross-curricular project weeks. The senior leadership team were very positive and enthused about how the project came together Friday. Kids have been buzzing, and I am a bit knackered still. I was just interviewed today again for a paper and website, and will send any press. In short, I believe the project was a success with lots of learning and linking[18]." When I spoke with Ramsing on the phone soon after we returned to the USA, he reported students were saying how the songs got stuck in their heads. Ramsing commented that he thought that catchiness of the music helped with retention of the academic lessons within the show.

Johnny Whiting, the UEL third-year performance student who played the role of Foss wrote the following in his blog. "After returning from the Christmas Holidays I was offered the opportunity to work with Beth Osnes, a Ph.D. professor of Theatre and Environmental studies at the University of Colorado, Boulder, the same university that I was able to attend during my semester abroad programme. Her project looked at performing an original musical constructed and performed by local youth that leads a community through the beginning phases of authoring a resilient city. Like a forum theatre format, this musical shows the environmental impact of fossil fuels beginning with the energy taken from the sun and shored within these fossils and concluding with human's engagement with fossil fuels such as the industrial revolution.

I was personally, honored to work with professor Osnes as it allows me an opportunity to understand the effective nature of theatre as a platform for change, similar to the works of Augusto Boal's Theatre of the Oppressed, whereby the audience themselves are asked to make changes and present new ways of engaging with issues. I really enjoyed working on this project and after the initial performance in the University of East London; we took the performance into a North London school [sic] in which we worked with the pupils in devising their own ways of helping make their city more resilient.

This was also a fine opportunity to return to script based work which I hadn't done since my semester abroad, it felt fantastic to return to working with a character and understanding a script in greater detail, I felt more open to play with and connect with the character and taking the performance techniques I had acquired during the university years into practice."[19]

NEW ORLEANS, LOUISIANA

Where: New Orleans
When: May 9, 2016
Community Partner: Saint Dominic's School
Short Description: This one-day residency at a Catholic Primary School highlighted the Catholic tradition of using theatre to disseminate new theological teachings. Fourth-graders at Saint Dominic's used performance to interpret the Pope's recent teaching related to climate change and related those teachings to their performance of *Shine*. This experience complimented the school's teaching of environmental stewardship (Fig. 3.7).

Description

The Catholic Church has a centuries-long history of utilizing theatre as a tool for disseminating theological teachings. Given the 2015 release by Pope Francis of his new theological teaching regarding climate change, *On Care for Our Common Home*,[20] the performance of *Shine* at Saint Dominic's Catholic School in New Orleans was both timely and in keeping with Catholic faith and practice. I was invited by Saint Dominic's principal Adrianne LeBlanc to do a one-day *Shine* residence. Together with three fourth-grade classes (68 students) we used performance to explore church teachings about our moral obligation to care

Fig. 3.7 Students at Saint Dominic's School responding to the question "Who wants to be the trilobite?". Photo by Rebekah Anderson

for our environment. We began at 8.30 a.m. in the gymnasium with all the costumes and art supplies readied to prepare for the all-school performance we would perform at 2.30 p.m. that day.

In a big circle in the gymnasium, we began with the warm up game Shaking Out Tension. Once I had their attention, I explained to them the Catholic Church's long tradition of using theatre to not only disseminate but to animate church teachings to the masses—people who likely weren't able to read theological texts. I asked the students if they had heard of the Pope's new encyclical, no one raised their hand, nor did they raise their hands when I asked if they imagined their grandmother or uncle had heard of it. I explained that we were going to use performance to convey the important lessons and teachings in this encyclical.

I described the story of the Corpus Christi and how the church used theatre to help spread this new teaching about the Eucharist to the people who at that time were largely illiterate. Some of these theatrical forms were performed on wagons and pulled through the streets to instruct and entertain people. I asked the students if this reminded them of any celebration in their own city, and they, of course, yelled out "Mardi Gras." I asked if they thought that this likely has its roots in the Medieval Christian drama, to which they replied yes.

Still assembled in our big circle, I announced that we were going to perform a prayer from the Pope's new teaching on caring for our Earth, a 16-line prayer at the conclusion of the encyclical for the students to dramatize. I numbered each line, printed out a copy, cut each line into a strip, and distributed one line to each of the student groups, made up of four to five students each. Each group was asked to create a dramatic statue or movement that communicated the meaning of the line. Groups were asked to assign one student to say the line aloud. Students were told that we were doing this in order to disseminate and animate the message and the spirit of the prayer. I gave them about three minutes to work on this. The three other teachers and I checked in with groups to see if they needed help or guidance. Once completed, I asked everyone to sit down with their groups in the circle. We began with the first line of the prayer, I walked around the room inviting each group to one at a time to stand, read, and enact their expression of the line. Some giggled while they did it, but all committed to the activity. The initial intention was to start the final performance for the school assembly with this prayer performed by the groups encircling the audience all around the outer edges of the gymnasium, but by the time we got to the actual performance, I was afraid there wasn't time, and, indeed, there wouldn't have been. If time had allowed, I think it would have been an effective way of communicating the connection between this theological text and the performance that followed for the entire school. As it was, this activity did seem to achieve this connection for the 68 fourth-graders. After doing this activity, I asked the students what was gained by acting out the lines instead of just reading them. One girl said it made it more exciting, and another said it helped them understand what the words really meant.

1. God of love, show us our place in this world
2. As channels of your love
3. For all the creatures of this earth,
4. For not one of them is forgotten in your sight.
5. Enlighten those who possess power and money
6. That they may avoid the sin of indifference,
7. That they may love the common good,

8. Advance the weak,
9. And care for this world in which we live.
10. The poor of the earth are crying out.
11. O Lord, seize us with your power and light,
12. Help us to protect all life,
13. To prepare for a better future,
14. For the coming of your Kingdom
15. Of justice, peace, love and beauty:
16. Praise be to you!

Amen.[21]

The fourth-grade teachers chose one boy and one girl to portray the two lead roles, Sol and Foss, who they thought would be sufficiently expressive and comfortable reading from the script in front of the entire school. We had nine students portray ancient plants wearing the green suits and three students portray ancient animals wearing the capes. They all entered from the back of the gymnasium and explored their way towards the sun standing center stage. Nearly all the students were excited to volunteer for each performance opportunity. Other students were given either Harvester or Fossil Fuel sashes.

When we got the point in the script where the Sun started referring to geological time to explain the evolution of the world, I was conscious of my uncertainly as to how this science meshed with what they taught the students in terms of the creation story and/or evolution. The kids, naturally, were thrilled to create a dinosaur that stomped across the stage and to enact the first humans making a fire. Seeing no visible signs of discomfort from the teachers present, we continued on through the script. When arranging for this residency, I had described the project thoroughly and had sent a script along to the principal, but had never explicitly asked about this issue. The scholar in me was interested to experience any reactions and negotiations through the process. When no objections emerged, I stopped focusing on this issue as we continued to progress. When I interviewed the fourth-grade religion teacher after the entire show, I explicitly asked her about this point, and she directly answered that they teach both science and religion. When I asked if there was any trigger in this show that seemed to go against church teachings, she said no, that there was nothing, all of it worked wonderfully and appropriately with what they taught. In fact, she said that when I was telling the kids about the Pope's teaching about environmental stewardship and how environmental degradation hurts those who are most vulnerable and did the least to cause the damage she could see the kids looking over at her and nodding, indicating that this was just what she had been teaching them. For her, this was a terrific way to reiterate what she was trying to impart to the students. She also added that she really liked the activity of having the students act out the various lines of a prayer and said she would like to try that out in the future to get them to understand the prayers more deeply.

After rehearsing Act One, students were asked to return their initial groups with whom they had expressed their line of the Pope's prayer. The opening prayer activity helped them be more prepared to author their own solutions for Act Two, since they had already worked together as a group to physically express something. It was also a useful scaffolding of the skills needed, as the first time they had the content of what they were expressing given to them to represent physically, and this time they had to author both the content and the representation of that expression. Once in their groups, I reminded them of what they should consider for effective communication:

Keep it local
Appeal to people's already held values
Focus on a single issue
Emphasize the positive
Identify co-benefits to climate and energy solutions
Frame solution as an opportunity

In designing the process for creating group solutions, special attention was given to inclusion—ensuring that each student's idea within each group would be heard and considered, not just the ideas of the most forthright students in each group. I asked each group to sit in a circle, close their eyes, and silently think of a solution they would share with their group. Then I asked each person to share their idea in just one sentence with their group, noting that the group would be choosing just one idea, but that they didn't have to decide yet. After each had shared, I asked the groups to consider if any common theme had emerged, noting if they saw a natural or organic way of synthesizing the solutions that were shared. Then I asked them to consider which solution might lend itself best to being acted out for the audience. I told them they were not making up a skit—just making a statue that conveyed their solutions in an active and interesting way—and that one person would narrate what the solutions were. To demonstrate, I had the teachers and I make a statue of our solution, which was a stay-cation (avoiding use of fossil fuels by staying home but being in the vacation mode). Two of us represented the mom and the dad, and one represented the little sister who received a new bike because of the money they saved on the travel expenses. One of the groups asked if they could combine two ideas, and I said sure, and encouraged them to try and link them in some way. Another group wanted to know if they had to be a still statue or if they could move, and I told them they could certainly have some movement in it. The students were given only about four minutes to identify a solution, create their statue or movement, and to decide who would narrate and what that spokesperson would say. Each group performed their solution for the rest of the groups and received feedback to make the expression of their idea more clearly communicated.

Throughout the day, the students remained engaged, likely because of the variety of tasks they were asked to do, such as decorating the long strips of paper to represent New Orleans, to decorate the flags with drawings of how New Orleans uses fossil fuels, or to make up their own dance for the final song. Although these students were well-behaved, in that they were responsive to their teachers and attentive to my facilitation, they relaxed into the dancing and let loose. During the final song *Shine*, students in their multiple lines were bumping hips, shimmying their shoulders, laughing, and grooving to the rhythms. Although there was a hint of misrule and abandonment to their rowdy dancing at the end, this contributed to the release and celebration that seems appropriate at the successful conclusion of a difficult task, such as rehearsing and performing a musical in a day. I had again brought the large colorful balloons to release at the end. The teachers wrote each group's solution on the balloons and held them up at the final song, to convey that they were supporting the ideas of the kids. It was decided by the teachers that they would release the balloons over the student's heads at the end. Again, like in London, the students nearly went wild reaching and diving for the balloons. Again, no one was hurt, but it certainly did release yet another level of inhibitions beyond the final song. It also seemed to impress and excite the students in the audience enormously.

Description of Youth-Authored Solutions

- Several of the groups dramatized how it saves on fossil fuels to ride a bike or walk instead of drive in a car—they also effectively acted out the co-benefits of this behavior. One group showed how two people walking looked so happy to be together, while the person driving looked sad, alone, and frustrated. Another student knelt down and crossed his arms in front of him with his head down, while the other students installed solar panels on his back as he portrayed a house. Other students portrayed two people eating at a nice restaurant and then sharing their leftovers with a poor hungry person they met along their way outside. One group had a girl lying on the ground facedown, undulating her body as a wave in a local lake, others were polluting her by dropping papers and their rubber bracelets on her. Once the narrator had described the problem of pollution, she described how those same people could also pick up that pollution, which they then enacted.

Lessons Learned and RecommendationsLessons Learned and Recommendations

1. Physical Performing Offers a Chance to Shine

After the performance, one of the fourth-grade teachers told me that in Act Two the girl who had portrayed the wave in the lake was extremely shy, and that she was surprised and impressed to see her so physically expressive, committed, and courageous. This example, coupled with the comment by the religion teacher that the opening prayer activity inspired her to want to incorporate more movement into her classroom, demonstrates that mounting a performance such as *Shine* can encourage teachers to consider integrating more movement into their instruction. This can result in providing alternative ways for students to express themselves with confidence, especially those students more comfortable with physical rather than verbal or written expression.

2. *Shine* Can Be Performed in a Day

Based on this mounting of *Shine*, it seems reasonable to expect that a teacher who has aptitude at leading youth in creative expression, or who has received training, could successfully facilitate youth in performing *Shine* in a day. The spoken delivery by the two primary characters could be improved by giving the two students a chance to read over the script a week or two before the day of the performance, allowing them more time to become familiar with the lines or even memorize them. If Sol and Foss read scripts, it is recommended that they *not* hold the script in front of their faces, so that their facial expressions can be seen by the audience.

3. Use Huge Balloons at Your Own Risk

The combination of dancing at the end of the performance and the release of huge balloons might just be too dangerous with kids. Even though the idea of it as realized at the University of East London seemed to be highly effective, with younger students, it seems to cause more mayhem than it does successfully convey its intended metaphoric message. Perhaps it could be effective for much smaller gatherings of children. The huge balloons are a hit and have the potential for engaging the entire audience in physical delight, but in practice they consistently created a frenzy that seemed likely to end in injury.

4. Connecting Faith with Climate and Resilience Planning

While reading *On Care for Our Common Home*, I kept wishing more people could benefit from how succinctly it made primary the relationship between humanity and the natural world. Unflinchingly, it addresses the many ways in which humanity seeks to dominate nature and how that has turned back upon us like a monster. I found the writing to be bold in its assertions, thorough in its scope, and soothing in many parts of its delivery. Reading this text made me feel truly optimistic about progress towards social justice and environmental stewardship that is at the heart of resilience planning. There are an estimated 1.2 billion Roman Catholics in the world,[22] if they could get even just some part of this theological text into their hearts and inspiring their actions, it could cause a tidal wave of change. But, of course, not many people read theological texts, even if those teachings are from the Pope. Pope Francis' leadership on issues surrounding climate—as one of the most prominent religious leaders—is what motivated me to partner with a Catholic community as part of this tour. This focus on faith leaders is highlighted in several effective strategies for climate communication. A guide created by ecoAmerica—*Connecting on Climate*—recommends aligning your climate communication with your audience's identity in terms of membership in a religious group, and aligning your message to the values that accompany that faith tradition.[23] Another effective strategy they recommend is to identify a messenger that your audience trusts, admires, and respects in order to significantly increase reception for your climate message.[24] Through the design of the residency at Saint Dominic's School, I explicitly sought to relate the content of *Shine* to the Pope's message. This focus on religious leadership on climate is supported by another publication by ecoAmerica, *Let's Talk Faith & Climate*, a practical guide for supporting discussions on climate change and faith among individuals and groups.[25] Since faith leaders are consistently included among the essential stakeholders in a city's planning for resilience, this guidance in navigating essential conversations is useful. Based on the success of this artist-in-residence at Saint Dominic's Catholic School, I recommend the use of *Shine* in faith-based schools.

5. The Connection to Early Performance Forms Comes Full Circle in Performance for Resilience

Also as a theatre scholar who teaches a graduate seminar on Global Ancient Theatre, the chance to integrate Medieval European theatre history into the New Orleans residency at Saint Dominic's Catholic School was satisfying. The study of theatre around the globe before Modernity is nearly all faith-based performance. The further back you go to the earliest accounts of performance by human communities, the more directly performance and faith become tied to recognition of the of natural forces that influence our lives. Performances were done to maintain balance with the natural world and to strengthen social bonds for the community's survival in the face of both natural and human-made threats. In fact, the farther back you go, the more the elements of ancient theatre resemble items on the agenda of a current-day resilience planning meeting. What we seek to achieve while seated at a conference table, ancient peoples sought to achieve through dance and enactment, investing their entire bodies into the effort, rousing all the evocative mediums available to them: music, movement, natural settings, and dramatic representation. This more holistic inclusion and acknowledgement of what we humans are made of (and, perhaps, of what truly motivates our actions) led to greater levels of balance with and connection to the natural world. While acknowledging that many aspects of our lives have changed over the years, if balance and connection are what we want to get back to as a human community, we may want to consider these early human methods and integrate embodied creative expression into our planning for resilience.

6. This One Was Personal

I was raised the youngest of ten children in a happy Catholic family, have an undergraduate degree in Theology from a Jesuit university, and have gone on to dedicate my life to the study and practice of performance for positive social change. Bethany Barratt in *Human Rights and Foreign Aid* eloquently writes, "by positive change I mean making the poorest people less poor and more empowered, making people who cannot express themselves free to do so, and making governments that would abuse the people to whom they are responsible unable to do so."[26] I would add to that description, making governments, corporations, and individuals that would abuse the natural world for profit or power unable to do so. Although I no longer practice as a Catholic, as a faith tradition it certainly nourished my early commitment to social equity and environmental stewardship, which is at the base of the *Shine* experience. This residency in New Orleans was deeply personal for me, stirring a lot of my own feelings about how my current involvement in the arts, academia, and in environmentalism converged in my first introduction to faith. The fact that this residency was in partnership with children—a time in my life when my own relationship with the Catholic Church felt most natural and clear—was especially healing and pleasurable.

I felt personally relieved and gratified that the religion teacher at Saint Dominic's found no conflict between what this show espouses and what the church teaches. If you are raised in a faith tradition, it is difficult not to continue seeking approval. This experience humbled me in knowing that I did care about positive affirmation from my faith heritage. Perhaps I even unconsciously came to this residency looking for it. I recommend that more communicators utilize their circles of community—including their faith traditions—as that is an effective method for climate communications and one for which useful tools exist.[27]

Feedback

During an interview, Emily Culotta, the religion teacher for the fourth-grade students reflected on her impression of the students' experience and how the performance related to what she teaches them. "I think that they enjoyed themselves so much, not only because they got out of class, but were able to come down here and interact with each other and learn a lot about resilience and what it takes for a city to be a resilient city. What I thought was cool too is that they got lessons that they have been learning in my class about being stewards of creation and about taking care of the Earth for the generations that are going to come behind them. They were all excited that they had just learned that and were able to bring it to life with today's play. I do think that it is something that is sort of overlooked these days, the care for our Earth. I really do think it is important for them to participate in these kinds of experiences to help them realize that the Earth that God has given them and has created for them is theirs to take care of. I honestly think that our youth are the best voice that we can hear. I know personally that I learn so much from just being with these kids. We one hundred percent incorporate science and faith and how they go hand in hand and how they are complimentary to each other. So, no, there is nothing today that as a religious teacher I was sitting saying, 'No, wait, we don't want them hearing anything like that.' It was great for them to have a different way, not just sitting in the classroom listening to one person lecture at them, to act it out and get into the meaning."[28]

MALOPE, SOUTH AFRICA

Where: Malope, South Africa
When: June 10, 2016
Community Partner: Malope Primary School, in Limpopo, the northernmost province in South Africa, Peace Corps
Short Description: Since the resilience concerns expressed by Malope Primary School administrators revolved around women's reproductive issues, specifically reducing stigma for menstruation, the core of this residency was a participatory play that dramatized the complete menstrual cycle. To advance their English-language skills, the fifth-, sixth-, and seventh-grade students rehearsed and performed the title song *Shine*. Female students created an additional skit on teenage pregnancy in anticipation of this residency that was focused on women's issues (Fig. 3.8).

Description

Twenty-five fifth-grade students stood at their desks and said in their best English, "Welcome teacher, how are you?" At Malope Primary School in South Africa—although the students' mother tongue is the Sepedi language (or Northern Sotho)—these students were all studying English. Our host, Claire Hackett who is a Peace Corps volunteer and their English teacher, had invited *Shine* for a one-day residence at

Fig. 3.8   A male student portraying the mother explains menstruation to a female student portraying the daughter. Photo by Melisande Osnes

their school. Hackett was one of the original volunteer collaborators who first traveled to the Navajo Nation with *Shine* when it was mounted with Tuba City High School. For this residency, Hackett's South African students from fifth-, sixth-, and seventh-grade English classes all participated in using performance as a tool for taking on resilience issues within their school, as a part of their resilience planning towards their community. Hackett had worked with these students in the weeks preceding my visit to memorize the *Shine* song as a method for learning English and as a segue into discussions about the concept of resilience planning. The girls in the seventh-grade class created their own skit as part of their Cool Girls Club at school around the subject of teenage pregnancy—an issue identified by them as important for the resilience of their own lives and their larger community.

On June 10, 2016 I, along with four US English-speaking volunteers (J.P., Melisande, Lerato Osnes, and Matthew Sanchez) and Hackett, led three similar sessions that lasted approximately 90 minutes with each of the grade levels. The school had identified that a major resilience issue for their community was that young girls who are menstruating often do not come to school during their period because they do not have access to menstrual pads and because of the stigma of menstruation. Hackett had previously attempted to partner with an organization that could provide reusable, washable pads, but they were unable to fulfill the request. Along with others in my home community of Boulder, we sewed enough sets of pads for the female members of the fifth, sixth-, and seventh-grade classes and brought them with us. Just to ensure that each female student would also have a pair of underwear to hold them in place, I also brought a pair of underwear for each female student. To ensure that the pads could be changed at school if necessary, we also brought a supply of plastic zip-lock bags for each girl along with a cloth carrying case for the plastic bags and the extra pad to ensure privacy. The principal of Malope Primary School, Sepheu Matsoge, had requested that we provide instruction for the proper use of these specific pads for the female students. Since the *Shine* experience is about using performance-based methods to engage youth in resilience issues, we worked with Matsoge in creating a participatory skit for reducing stigma associated with menstruation and for explaining how these pads could be used by the students.

Once we were introduced to the class by Hackett, we formed a circle with the students and did two activities: Name and Gesture and Shaking Out Tension. In each of the three classes, when volunteers were requested to play the role of the mother and the daughter in the skit we prepared, a male student volunteered to play one of the roles. A female student was asked to volunteer to play the other part so there would be gender equity in the major roles. Previous to the session we had used red, one inch-wide masking tape to create a large outline of the female reproductive organs in the center of the floor, which included the ovaries, the fallopian tubes, the uterus, the cervix, and the vagina. This diagram nearly took up the entire open space in the classroom. Since the students were all in their school uniforms, we wrapped the mother and her daughter characters in a piece of traditional South African fabric to represent a skirt worn at home. Their opening lines were as follows:

Daughter: Mother, my breasts are feeling so tender today. I feel cramps down here that hurt. And my little brother is so irritating; I could throw him out the door. But Mother, the worst thing of all is that I am bleeding from between my legs. Mother, am I going to die?

Mother: No, my daughter. What you are experiencing is called menstruation. This is exciting.

Daughter: I don't feel excited. I feel miserable.

Mother: You are becoming a woman so your body is changing. How about you and I take a tour through your body so you can understand what is happening inside.

At this point we asked for two volunteers to be eggs and each stand in one ovary. We asked for two other volunteers to be the uterine walls, handed them each two red scarves (one for each hand) and asked them to stand along either side within the uterus. We asked everyone remaining to form a circle around the female reproductive system. We had a poster of the female reproductive system leaning on the chalkboard and the names of each part represented in our drawing written on the board. Hackett reviewed each of the terms with her students and pointed out each term on the poster and on the floor diagram, asking the entire class to repeat each term aloud as it was introduced. Then I announced that ovulation was about to occur which would start the cycle of menstruation. The students in the surrounding circle were asked to turn to their left and march in a circle while repeating the rhythmic chant "oh, ah, ah, oh, oh, ah–ovulation, oh, ah, ah, oh, oh, ah–ovulation" until the designated egg had slowly made its way from the ovary, through the fallopian tube, and into the uterus. Next the uterine walls were cued to get excited, wave their red scarves and prepare a lovely home around the egg in preparation for possible insemination. At this, students waved their red scarves and raised their arms around the student representing the egg. Then I told everyone to stop and wait. We waited together in silent anticipation. I asked what we were waiting for. One student responded we were waiting for the male seed to make a baby. Hackett wrote the term "sperm" on the chalkboard and asked the students to repeat it aloud together. Since no sperm was arriving, we instructed the lining of the uterus to shed and flow, along with the egg, through the small opening in the cervix and travel out of the body through the vagina. At this the two students representing the uterine walls and that egg were asked to walk out of the diagram through the cervix and the vagina. I took the red scarves from those students and asked for two more volunteers to represent the uterine walls and stand on either side of the uterus. That, I announced, was one complete menstrual cycle.

We prepared to do another cycle. This time everyone in the circle was asked to turn to the right and march in a circle while repeating the rhythmic chant "oh, ah, ah, oh, oh, ah–ovulation, oh, ah, ah, oh, oh, ah–ovulation" until the egg from the other ovary slowly made its way from the ovary, through the fallopian tube, and into the uterus. Again, the uterine walls were cued to get excited, wave their red scarves and prepare a lovely home around the egg in preparation for possible insemination. At this, students waved their red scarves and raised their arms around the student representing the egg. Again, everyone stopped and waited in silent anticipation. This time I announced that we would experience what happens when a male sperm does make it to the egg. Hackett portrayed the male sperm and walked through the vagina and cervix into the uterus and linked elbows with the student portraying the egg. At this they squatted down as small as possible together with the uterine walls holding their arms around them protectively. The students in the surrounding circle were instructed to gesture one arm from outside the circle and gradually move it towards the center while they chanted "Grow, grow, grow, one month. Grow, grow, grow, two months. Grow, grow, grow, three months", and so on until reaching nine months, at which point the egg and sperm were fully grown and standing tall. Next, the surrounding circle put their arms around each other and helped squeeze the baby out through the dilating cervix and the vagina. Hackett and the student, arms still linked, made their way through the birth canal portraying great struggle. Once they were born and Hackett let out a cry like a newborn baby, the class erupted in applause.

At this point we returned to the daughter and mother.

Mother: My dear, now do you understand what is happening with your body during menstruation?

Daughter: Yes, mother, thank you. I understand.

Mother: Very good. Now, my dear, go get ready so you can go to school.

Daughter: What? Go to school? I'm not going to school with my period. It might smell. It might bleed through my uniform. No way.

Mother: Not to worry, my dear. I have some reusable pads for you to use. (She gets the pads that we had brought and a pair of underwear. She demonstrates how to secure it on the underwear, how it works, how it can be changed if soaked with blood, and how it could be discretely carried to and from the toilet in the cloth bag.)

Daughter: Oh mother, this is wonderful. Now I will not fall behind in my lessons at school.

Mother: No, my dear. If you are going to be a leader in our community to help make it resilient, you must continue your schooling.

Daughter: Yes, mother. I want to do that.

As a conclusion, Hackett led all of the students in the following call and response:

Hackett: Even though I'm a girl.
Class: Even though I'm a girl.
Hackett: And even though I have my period.
Class: And even though I have my period.
Hackett: I CAN STILL GO TO SCHOOL!
Class: I CAN STILL GO TO SCHOOL!

We intentionally included the male students in this entire class so that they too could learn about the female reproductive system and play a part in helping to reduce the stigma of menstruation, both literally and figuratively. It was remarkable that the male students volunteered to play female roles of the mother or the daughter, but also that they delivered the lines with conviction and what appeared to be authentic concern. During the part of the skit when the use of the reusable pads was being demonstrated, each student in the class was invited to touch the waterproof fabric that kept the blood from soaking through into the underwear. In each class, all of the students took the offer to feel the fabric when it was offered to them. Throughout the entire skit, students participated, kept their focus on the action, and exhibited excitement at the outcomes of the various actions.

Near the end of the school day, all of the participating students assembled in the schoolyard to sing and perform their choreography of the title song *Shine*. Dust rose from their stomping, girls' skirts flew from the twirls during the line, "turn around, touch the ground" and energy filled the school grounds, with an audience of students, teachers, principal, and our volunteer team cheering them on. At the conclusion, we went into one of the larger classrooms for the performance of their youth-authored solution about the central resilience theme they had identified—teenage pregnancy.

Description of Youth-Authored Solutions

In conjunction with this one-day *Shine* residency at Malope Primary School, the seventh-grade girls in the Cool Girls Club wrote and performed a play as their youth-authored solution. The girls in the club portrayed all the various roles in the small space at the front of the crowded classroom. The play centered on a girl named Lerato who became intimate with a boy she liked. On a date, she kissed him, which led to more sexual activity. Her friends reported her behavior to the teacher, but Lerato denied it. Even though they jeered at her and accused her of having had sex, she denied their claims as rubbish. The boy appeared and outed her as having been sexually active with him. Later she stopped getting her period and suspected she may have become pregnant. In another scene, she was lamenting the fact of her pregnancy to her family. Her mother tried to convince the young man to marry her daughter, a request that his parents resisted. As a conclusion, the performers sang a song they had composed on the subject and story of their title character, Lerato. Their play concluded with a call for abstaining from sexual activity.

Directly following the play—with the audience still present—Hackett challenged the entire gathering of male and female students to suggest actual lines the character Lerato could say to the boy if he pressured her for sex. This, in turn, prompted two of the local female teachers to likewise comment on the responsibility of the boys in these kinds of situations. One female student from the audience suggested that she should say if a boy really loved her, he wouldn't press for sex. A male student suggested she say that if the boy in the skit wasn't ready for the responsibility of fathering a child, he wasn't ready for sexual relations.

Due to scheduling differences for each of the classes, the session that we did with the sixth-grade students was 30 minutes longer than the other sessions, which afforded the time to work with this class to create a Statue Park of Guiding Values. In the field behind the school, we got into one large circle and did the activity Zip, Zap, Grr to gather the students' focus and energy. Still standing in a circle, the idea of a Statue Park of Guiding Values was presented. Students were asked if they have ever seen a statue in a city center that represents the values of that community. A few described statues they had seen and what values they thought these statues may have represented. We divided the class into groups, each with one of our English-speaking volunteers, and challenged each group to identify a guiding value for their community and then make a statue that expressed that value. The results are as follows:

- The first group had the adult volunteer in the group with his fist raised leaning over someone smaller and the rest of the group attempting to hold him back. While they held the pose, the rest of us walked around their statue until someone guessed the value: do no harm.
- The second group had movement as part of their statue. One person had her back to the rest of the group and, keeping her body straight, fell backwards into the waiting arms of the group, which represented trust.
- The third group represented support by standing in a circle with interlocking arms, leaning back, and supporting each other.
- The last group represented respect by posing as a class listening respectfully to their teacher.

Lessons Learned and Recommendations
1. Be Responsive to Each Community

This part of the tour was not tied to the formally scripted story of *Shine* but still focused on the theme of youth-authored resilience planning and the use of music from *Shine* to invigorate the creative process. The alternative arrangement of this artistic residency allowed the performance-based methods to more fully explore issues of resilience identified by the community itself, such as the stigma of menstruation and the challenges of teenage pregnancy. Although the participatory play on menstruation was much more simplistic and rough than the story of *Shine*, it heartily captured the interest and involvement of both the male and the female students. For this skit, I borrowed from a street theatre performance I had co-created with students for a maternal health organization in Guatemala. We modified the skit slightly to include modeling the use of reusable menstrual pads.

This residency in South Africa makes evident that there are a variety of ways that participatory performance can be used to engage youth in city planning for resilience. The story of *Shine* focuses primarily on energy and climate, while this residency focused more heavily on women's issues. Artistic achievement in performance can often be increased when the subject of a performance is determined ahead of time; narratives can be carefully constructed, songs composed, and danced choreographed. However, there can be payoffs from less highly produced participatory performance work as well, chief among them that they can be more responsive to the community's self-identified resilience issues, and they can have an immediacy to them that can be highly engaging. The highly produced option can be effective for public performances, as the increased artistic achievement may enhance audience receptivity to youth messages. However, the more immediate and rougher option that more directly represents priorities from within the community may also be engaging and, perhaps, more relevant for public performance. Both approaches can support youth and community engagement, simply in different ways. Options along this range can be balanced to maximize the benefits from youth achieving artistic excellence with the benefits of youth having maximum agency in authorship around that subject.

2. Challenging Youth-Authored Solutions

When the Cool Girls Club concluded their performance about Lerato, they left the blame for the unwanted pregnancy at Lerato's feet and the responsibility for the child with her and her family. The irony is that this skit was authored by young women, yet it unfairly disadvantaged the female position in the situation of an unwanted pregnancy. Youth solutions, like solutions put forth by any constituency, can be improved upon through critical reflection and by asking questions that expose underlying assumptions and biases. In this case, it seemed appropriate that the questions originated from the girls' own teachers since they are a part of the female students' own community. It also seemed highly effective that the teachers took on the role of the facilitators for the post-performance conversation. In addition, it was highly effective that they led by asking questions of the youth rather than providing answers or correcting them.

CONNECTICUT, USA

Where: Connecticut and New York
When: June 30, 2016
Community Partners: Manhattanville College, Purchase, NY
Short Description: In order to create a professional video recording of *Shine* that captured Fredric's direction and choreography, Fredric arranged for a performance near his hometown in Connecticut. After rehearsing with area youth for an entire week, we arranged for a multi-camera shoot of several performances at nearby Manhattanville College (Fig. 3.9).

Fig. 3.9 Foss and his followers dancing. Photo by ©2016 Steven Sutton, DUOMO

Description

From Arthur Fredric's hometown in Connecticut, we arranged for a performance of *Shine* with the goal of professionally documenting the performance, complete with Fredric's choreography and stage direction. This was not arranged as a public event, but, rather, solely as a video and photographic shoot, such that only a small invited audience was present. After Fredric and Lisa Denton rehearsed the show intensively for a week with primarily community youth, we mounted the show on the main stage of nearby Manhattanville College. Cinematographer Michael Castaldo directed a crew of cameramen to operate one crane-operated camera to follow the action, two side cameras for close ups, and a GoPro camera fixed directly above the performance area to capture an aerial-view of Fredric's choreography and staging. Steven Sutton, who is a long-time friend of Fredric and a highly acclaimed professional photographer, donated his time to photograph both a rehearsal and the two performances of *Shine* performed one right after the other on June 30, 2016. Meridith Richter, a Technology, Media, and the Arts Major at CU edited the footage to create the full performance video recording available at https://vimeo.com/194833723.

Tom Wasinger was also present in Connecticut, reprising what he had arranged for the New York City performance, along with National Dance Institute musician, Tim Harrison. J.P. Osnes designed the lights and served as production manager. A recent theatre graduate of Manhattanville College, Megan Correia, performed as Sol, and Emmett Smith reprised his performance of Foss from the New York City performance. Jerusha Wright of Sandy Hook, performed as the Seed Sower. The rest of the cast of 19, ranging in age from nine to twenty-one, and were area youth who had previously performed in the Danbury Music Centre Nutcracker Ballet, for which Fredric has been the artistic director for nearly two decades. One of the hallmarks of this yearly production is their radical inclusiveness of all levels of experience and aptitude. Our cast for *Shine* was made up of youth from several surrounding towns and cities, including: Newtown, Brookfield, Danbury, Carmel, New Fairfield, New York, and Sandy Hook. Three of the young performers from Sandy Hook in our performance had lived through the school shooting that occurred there in 2012. Participating in the Nutcracker in the time following this tragedy was instrumental for them returning to normality in their lives. Fredric and Denton mindfully used the arts for community healing—wisely having therapists and even therapy dogs at rehearsals through the difficult times that ensued.

During the rehearsal process for *Shine*, Fredric taught the dances and the staging, and Denton demonstrated combinations and worked with various performers who benefitted from direct instruction. She also worked with the performers of Sol and Foss to rehearse their dramatic performance. Chase Sutton served as Fredric's assistant choreographer. Rehearsals lasted the full day, with the production team working well into the night to arrange all of the other details for the shooting. On the evening before the video shoot, Joshua Sperling, who is Fredric's nephew and one of the primary scientific collaborators for this project, arrived with a fellow scientist, Camron Adibi, who assisted Sperling in working with the youth to connect energy and climate science to their performance experience during the final rehearsal the morning of the shooting. As he had for the New York City performance, Sperling again willingly put on his full-black Lycra body suit to serve with me as a stagehand in the performance. During the production week, all of us behind the scenes enjoyed the hospitality of Fredric and Denton's home for multiple sleepovers during which more cooking, conversation, and preparations occurred than sleeping.

Description of Youth-Authored Solutions

- The first group looked at the problem of not enough electricity for the future. Their solution was a bike that had a battery attached to it, so it could store power when pedaled. Then, when you got home you could plug it into a socket that could charge a battery as well as provide power for the community.
- The second group looked at the problem of car emissions and proposed the solution of a hydro-powered car.
- The third group looked at the problem of global warming. Their solution was to look for alternate sources of energy, such as putting solar panels on the top of most buildings.
- The fourth group looked at the issue of animal extinction. Their solution was to raise awareness by setting up billboards, making commercials, and change our daily habits, such as using reusable bags when shopping.

BOULDER, COLORADO

Where: Boulder, Colorado
When: July 11–15, 2016
Community Partner: CU Science Discovery
Short Description: CU Science Discovery offered a week-long camp that involved performance-based methods to engage youth in energy and climate science. Their culminating performance of *Shine* featured a 6-foot long paper dinosaur and a geological timeline highlighting various periods from the script. The addition of Art, converted the education of STEM (Science, Technology, Engineering, and Math) to STEAM (Fig. 3.10).

Description

The geological timeline created by participants of the Science Discovery camp for *Shine* was a work of art. Stretching over 2 meters in length, it began in the Carboniferous Period, traveling through the Triassic Period, the Industrial Revolution, and ending in the current day. One student said when pointing to the clouds above the present day, "The future is bright if we make it bright." A CU student, Mariel Kramer, assisted by a Science Discovery teaching assistant Rachel Sharpe,

Fig. 3.10   Geological timeline. Photo by Meridith Richter

led nine youth ranging in age from nine to twelve in a week-long, all-day camp on the CU campus. CU Science Discovery offers a variety of hands-on STEM camps for kids ages five to eighteen. They were receptive to a camp that would involve performance-based methods to engage youth in energy and climate science. A performance for the parents of participants and the general public was planned for the afternoon of the concluding Friday class. It should be noted that the leader of this week, Kramer, had participated as an audience member in the Boulder SEEC performance of *Shine* in October of 2015.

Each morning began with Kramer engaging participants in theatre-based games and activities in the grass, before convening in their designated classroom in the Earth Sciences building on the CU campus. Two of the participants, Lerato Osnes and Finella Guy, had rehearsed in the weeks previous to the camp to perform the two lead roles of Sol and Foss respectively. On Monday, Kramer engaged the youth in first researching and then painting a geological timeline of the years spanned by the action of the play. The youth threw themselves into this task with vigor and were careful to note that the dinosaurs actually came after the Carboniferous Period and, therefore, are not what fossil fuels are made of. They were thrilled to note that the gas station Sinclair got it wrong by having a dinosaur on their logo. Painting the illustrations for each period led them to explore what happened in each era as it related to the formation of fossil fuels and the ways in which humans began to impact climate with their overuse of fossil fuels during the Industrial Revolution. This artistic activity, and those that followed, provided the context in which the science behind the story became emphasized, which was in keeping with this being a STEM camp.

On Tuesday, the entire group met in the morning at the National Center for Atmospheric Research (NCAR), which is nestled against the foothills to the Rocky Mountains just a few miles from the CU campus. There they explored the paths near NCAR, toured the exhibit on energy formation and climate, and visited the interactive exhibits and learning labs for youth. Back on the CU campus, they commenced on Wednesday at the beginning of the script, and worked their way through, doing the corresponding artistic activities, such as decorating the strips of paper and decorating the fossil fuel flags, that matched each song. By the end of the day on Thursday, all of the songs had been learned and choreographed, and their solution skit had been created. Friday morning was spent running the entire show a few times and co-ordinating with the many props the youth had created for the performance, including a huge dinosaur made out of green paper that required three students to manipulate.

As guests entered the performance on Friday afternoon, performers explained the significance of the geological timeline to the story they were about to witness. One of the highlights of the Friday afternoon performance was that it took place in an open area of the Earth Science Library, directly in front of a giant cast of a life-size Stegosaurus Stenops that is affixed to the wall. This is a replica of one of the most complete skeletons ever found, that happens to be from nearby Cañon City, Colorado. An audience of approximately 50 family members, other camp attendees, and camp staff witnessed their performance. Since it was a small cast, the performers asked for volunteers from the audience to help with the weaving of the fabric of community. The scheduling of the performance was convenient for the parents to pick up their kids, but, unfortunately, did not leave time for the youth to reflect on the performance experience as a group afterwards.

Fig. 3.11 Enacting a scene about a local store where the currency is green carbon credits for Act Two. Photo by Meridith Richter

Description of Youth-Authored Solutions

This group decided to create just one skit all together for Act Two. Kramer performed as the owner of a local store where the currency was green carbon credits (Fig. 3.11). A young man approached the checkout with his vegetables.

Clerk: Wow, I see you have a lot of vegetables there.

Customer Number One: Yeah, I'm a vegetarian.

Clerk: I bet you saved so many CO_2 and methane emissions because of your dietary choices. Congratulations, (swipes his card) you get **50 green points.**

(The next customer approached and mimed placing a purchase on the counter.)

Customer Number Two: I'm buying this local slab of meat.

Clerk: Congratulations, you're saving many CO_2 emissions that it would have taken to transport that meat here.

(Another girl approached the counter and thrusts flip flop shoes in front of the clerk's face.)

Seller: You wanna buy these? I made them on my farm from all organic materials.

(The clerk proceeds to buy her wares by adding green credit to her card. The next customer, a vegan, was highly congratulated for her lifestyle choice and was awarded 100 green points to her card.)

Lessons Learned and Recommendations
1. A Week-Long Camp Is a Perfect Match

A week-long, all day camp, is an ideal format for thoroughly exploring the science and the artistic expression of *Shine*. This camp allowed for the additional creation of the timeline, which grounded the performance experience. Both the schedule and the physical setting through Science Discovery allowed for ample time outdoors to immerse the experience in the natural world, which significantly reinforced the ecological message permeating the show. The relaxed summer pace made for a pleasant experience for both the youth performers and the leaders. The offering of this camp allowed youth with an artistic sensibility to increase their scientific knowledge, while simultaneously experiencing a wide range of artistic mediums in which to be expressive.

2. Adding Arts to STEM Is Gaining STEAM

Recently, the idea of converting the education of STEM (Science, Technology, Engineering, and Math) to STEAM (Science, Technology, Engineering, Arts, and Mathematics) has been gaining momentum.[29] Some educators suggest that we shape STEAM programs by exploring opportunities where art naturally fits in the STEM arena, where art can be treated as an applied subject—just like math and science. This works well with *Shine*, given that the basis of this work is in the field of applied theatre. There is a practical goal associated with applied theatre, most often expressed as a social concern identified by the community hosting the workshop or performance. The goal of *Shine* is not so much for the arts to serve as merely a communication device for energy and climate science, but rather to integrate both art and science to more accurately represent the global position we find ourselves in now—where our excessive use of energy has led to climate change. Given that the solutions to this conundrum are not solely scientific, but also social and behavioral, the combined use of performance and science seems exceedingly appropriate. It's not likely that we will solve the challenges associated with climate change with scientific innovation alone. It will take our collective global population authoring and enacting a new story around the amount of energy we consume and how we produce that energy.

Feedback

Mariel Kramer shared the following feedback. "The weeklong structure of the camp allowed for plenty of in-depth study of the concepts introduced in *Shine*, such as the study of geologic time, photosynthesis, basic evolutionary biology, ecology (primarily the carbon cycle,) and the greenhouse effect. The ability for students to alternate between these scientific topics and engaging visual and performance art projects, that all built up to the final performance, kept them engaged throughout the entire day, and excited about learning new things. I made a point to continually reinforce the scientific purpose for the inclusion of each piece of script, asking students to recall whichever concept a given line was referencing. Through this dynamic form of learning, where we covered a scientific concept and the students then expressed their understanding through dance, song, drama or visual art, the merits of a STEAM-based curriculum were clearly demonstrated. I had anticipated the students would become quickly bored, as I planned to get 'into the weeds' of some of the topics discussed, but the students remained impressively engaged with the material, often introducing topics for discussion and creative ideas for our upcoming performance unsolicited. Our time allotted throughout the week allowed ample room for students to complete all props by hand, including a beautiful geologic timeline complete with appropriate illustrations for each period and a paper dinosaur that was big enough to be carried by three students, to cover material on each topic mentioned in the script in depth, develop our own full-cast skit to be included as the finale, rehearse our performance thoroughly, and spend a day hiking and exploring NCAR's museum in order to engage with the material through a different lens."

A few bits of feedback I find will be helpful for future productions of the show.

1. Keep the dynamic structure of the rehearsal period: alternating between what could be considered a more traditional style of teaching scientific-concepts (lecture-based,) an artistic expression of concepts while they are still fresh in the students minds (i.e., now that you have learned all about the carbon cycle, can you paint me a model of what you think it might look like?) rehearsal time, and improvise games (indoor or outdoor, weather dependent) to ensure that all of the students are comfortable performing and acting in front of one another, a task that can be scary to the inexperienced performer.

2. Ensure a backup plan is in place for inclement weather! We were very excited to be able to "book" access to a bandshell on CU's campus that has a stunning view of the mountains for the performance, but unfortunately were unable to perform there due to a thunderstorm. This led to a bit of scrambling and confusion, as we did not establish a hard-and-fast back-up plan in the case of rain. If future performances are planned to be outside, make sure to think of great alternatives for backup as well. We were very lucky to be able to host the show in the Earth Sciences Library at the University, which ended up being the perfect setting for the show.

3. Think of all the ways to limit waste and recycle materials when making props for the show. For instance, it was exciting that we were able to use stuffing from old pillows to represent our "carbon" during the Industrial Revolution portion of the show. We simply dabbed the stuffing with black paint and allowed it to dry. The use of the pillow stuffing allowed for easy cleanup, with very little mess leftover, and the ability to reuse the stuffing again for future performances."[30]

NOTES

1. Beth Osnes, Adrian Manygoats, and Lindsay Weitkamp, "A Framework for Engaging Navajo Women in Clean Energy Development through Applied Theatre," *Research in Drama Education: The Journal of Applied Theatre and Performance* 20, no. 2 (April 3, 2015): 242–257, doi:10.1080/13569783.2015.1019445; Beth Osnes, "Engaging Women's Voices through Theatre for Energy Development," *Renewable Energy*, no. 0 (2012), doi:10.1016/j.renene.2012.01.036; Beth Osnes, *Theatre for Women's Participation in Sustainable Development - Routledge* (New York, NY: Routledge, 2014), http://www.routledge.com/articles/theatre_for_womens_participation_in_sustainable_development/; Barbara Farhar et al., "Engaging Women in Clean Energy Solutions Workshop at the World Renewable Energy Forum (WREF2012)," *Gender & Development* 20, no. 3 (November 2012): 616–17.

2. Eagle Energy, "Navajo Solar," 2013, http://elephantenergy.org/Navajo_Solar.html.

3. Osnes, Manygoats, and Weitkamp, "A Framework for Engaging Navajo Women in Clean Energy Development through Applied Theatre."

4. Greg Guibert, "City of Boulder Resilience Strategy" (Boulder Colorado: City of Boulder, 2016), [CSL STYLE ERROR: reference with no printed form.].

5. James White, "The Show," June 14, 2015.

6. Cathy Deely, "Performance," June 15, 2015.

7. Michelle Fox, Marcus Moench, and Rachel Norton, *Beyond Resilience* (Institute for Social and Environmental Transition-International, 2015), 6–9.

8. Leela Stoede, Leela Stoede Interview, October 12, 2015.

9. Charlotte Gerrity, Charlotte Gerrity Interview, October 12, 2015.

10. Lucinda Stewart, Lucinda Stewart Interview, October 12, 2015.

11. Brett KenCairn, "Youth Performance," October 7, 2015.

12. Phaedra Pezzullo, "SEEC Performance," October 4, 2015.

13. Shelly Sommer, "Bravo on Performance," October 24, 2015.

14. Citiscope, "What Is Habitat III?," *What Is Habitat III?*, 2016, http://citiscope.org/habitatIII/explainer/2016/09/what-habitat-iii.

15. United Nations, "Children and Youth Assembly," *Habitat III*, October 15, 2016, https://habitat3.org/programme/children-and-youth-assembly/.

16. Morana Stipisic, "Performance at The City We Need," October 28, 2016.

17. Augusto Boal, *Theatre of the Oppressed* (New York: Theatre Communications Group, 1985), 112.

18. Soren Ramsing, "Follow Up On Shine in London," January 29, 2016.

19. Johnny Whiting, "Shine- Working with Beth Osnes," *Johnny Whiting Public Project Blog*, February 26, 2016, https://johnnywhiting.wordpress.com/page/5/.

20. Pope Francis, *Encyclical Letter Laudato Si' of the Holly Father Francis, On Care for Our Common Home* (Vatican City: Vatican Press, 2015), http://w2.vatican.va/content/dam/francesco/pdf/encyclicals/documents/papa-francesco_20150524_enciclica-laudato-si_en.pdf.

21. Ibid., 180.

22. BBC News, "How Many Roman Catholics Are There in the World?" (BBC, March 14, 2013), http://www.bbc.com/news/world-21443313.

23. Ezra Markowitz, Caroline Hodge, and Gabriel Harp, "Connecting on Climate: A Guide to Effective Climate Change Communication" (New York: Center for Research on Environmental Decisions, Columbia University, 2014), 9–10.

24. Ibid., 10.

25. Kirra Krygsman and Meighen Speiser, *Let's Talk Faith & Climate: Communication Guidance for Faith Leaders* (Washington, D.C.: ecoAmerica, 2016).

26. Bethany Barratt, *Human Rights and Foreign Aid: For Love or Money?* (London; New York, NY: Routledge, 2008), 2.

27. Krygsman and Speiser, *Let's Talk Faith & Climate: Communication Guidance for Faith Leaders.*
28. Emily Culotta, Emily Culotta Interivew, May 9, 2016.
29. Anne Jolly, "STEM vs. STEAM Do the Arts Belong?," *Education Week Teacher*, November 18, 2014.
30. Mariel Kramer, Feedback about Shine Experience, May 20, 2017.

REFERENCES

Barratt, Bethany. *Human Rights and Foreign Aid: For Love or Money?* London; New York, NY: Routledge, 2008.

BBC News. "How Many Roman Catholics Are There in the World?" BBC, March 14, 2013. http://www.bbc.com/news/world-21443313.

Boal, Augusto. *Theatre of the Oppressed.* New York: Theatre Communications Group, 1985.

Citiscope. "What Is Habitat III?" *What Is Habitat III?*, 2016. http://citiscope.org/habitatIII/explainer/2016/09/what-habitat-iii.

Culotta, Emily. Emily Culotta Interivew, May 9, 2016.

Deely, Cathy. "Performance," June 15, 2015.

Eagle Energy. "Navajo Solar," 2013. http://elephantenergy.org/Navajo_Solar.html.

Farhar, Barbara, Beth Osnes, Priyadarshini Karve, Long Seng To, and Nicole Speer. "Engaging Women in Clean Energy Solutions Workshop at the World Renewable Energy Forum (WREF2012)." *Gender & Development* 20, no. 3 (November 2012): 616–17.

Fox, Michelle, Marcus Moench, and Rachel Norton. *Beyond Resilience.* Institute for Social and Environmental Transition-International, 2015.

Gerrity, Charlotte. Charlotte Gerrity Interview, October 12, 2015.

Guibert, Greg. "City of Boulder Resilience Strategy." Boulder Colorado: City of Boulder, 2016. [CSL STYLE ERROR: reference with no printed form.].

Jolly, Anne. "STEM vs. STEAM Do the Arts Belong?" *Education Week Teacher*, November 18, 2014.

KenCairn, Brett. "Youth Performance," October 7, 2015.

Kramer, Mariel. Feedback about Shine Experience, May 20, 2017.

Krygsman, Kirra, and Meighen Speiser. *Let's Talk Faith & Climate: Communication Guidance for Faith Leaders.* Washington, D.C.: ecoAmerica, 2016.

Markowitz, Ezra, Caroline Hodge, and Gabriel Harp. "Connecting on Climate: A Guide to Effective Climate Change Communication." New York: Center for Research on Environmental Decisions, Columbia University, 2014.

Osnes, Beth. "Engaging Women's Voices through Theatre for Energy Development." *Renewable Energy*, no. 0 (2012). doi:10.1016/j.renene.2012.01.036.

———. *Theatre for Women's Participation in Sustainable Development - Routledge*. New York, NY: Routledge, 2014. http://www.routledge.com/articles/theatre_for_womens_participation_in_sustainable_development/.

Osnes, Beth, Adrian Manygoats, and Lindsay Weitkamp. "A Framework for Engaging Navajo Women in Clean Energy Development through Applied Theatre." *Research in Drama Education: The Journal of Applied Theatre and Performance* 20, no. 2 (April 3, 2015): 242–57. doi:10.1080/13569783.2015.1019445.

Pezzullo, Phaedra. "SEEC Performance," October 4, 2015.

Pope Francis. *Encyclical Letter Laudato Si' of the Holly Father Francis, On Care for Our Common Home*. Vatican City: Vatican Press, 2015. http://w2.vatican.va/content/dam/francesco/pdf/encyclicals/documents/papa-francesco_20150524_enciclica-laudato-si_en.pdf.

Ramsing, Soren. "Follow Up On Shine in London," January 29, 2016.

Sommer, Shelly. "Bravo on Performance," October 24, 2015.

Stewart, Lucinda. Lucinda Stewart Interview, October 12, 2015.

Stipisic, Morana. "Performance at The City We Need," October 28, 2016.

Stoede, Leela. Leela Stoede Interview, October 12, 2015.

United Nations. "Children and Youth Assembly." *Habitat III*, October 15, 2016. https://habitat3.org/programme/children-and-youth-assembly/.

White, James. "The Show," June 14, 2015.

Whiting, Johnny. "Shine- Working with Beth Osnes." *Johnny Whiting Public Project Blog*, February 26, 2016. https://johnnywhiting.wordpress.com/page/5/.

CHAPTER 4

Conclusion

Abstract In addition to the distinct lessons learned from each performance of the tour—shared in Chap. 3—many themes emerged as a result of surveying the experience of the tour as a whole. My initial suspicion and now conviction, is that including youth in resilience planning is both critical for youth—the community leaders of tomorrow—and for our communities, who benefit from the joy, fresh perspective, and sheer energy boost that youth participation provides. This chapter looks at specific aspects of the performance experience to make the case for why youth inclusion in resiliency planning is important for youth, why it is important for our wider communities, and why performance itself is uniquely effective in addressing our resilience challenges.

Keywords 100 Resilient Cities · Youth engagement · Applied theatre Climate change · Energy · Resilience · Inclusion · Resilience narratives · Play · Performance · Social capital · Dramatic metaphor Celebration

In addition to the distinct lessons learned from each performance of the tour—shared in Chap. 3—many themes emerged as a result of surveying the experience of the tour as a whole. My initial suspicion and now conviction, is that including youth in resilience planning is both critical for youth—the community leaders of tomorrow—and for our communities, who benefit from the joy, fresh perspective, and sheer energy boost

B. Osnes, *Performance for Resilience*,
DOI 10.1007/978-3-319-67289-2_4

that youth participation provides. This chapter looks at specific aspects of the performance experience to make the case for why youth inclusion in resiliency planning is important for youth, why it is important for our wider communities, and why performance itself is uniquely effective in addressing our resilience challenges. To begin with, I examine the power of performance, as revealed and/or confirmed by the tour of *Shine* (Fig. 4.1).

CONTRIBUTING TO YOUTH

Hope Is Vital

It is essential to discuss climate concerns with youth in a positive way. Hope and possibility must be our baseline. I use the word "hope" in the sense of believing in the chance that something good could happen and confidence in the possibility of some form of a positive outcome to our current crises. Michale Rohd, author of *Theatre for Community Conflict and Dialogue: The Hope is Vital Training Manual* writes that, "The act of expression is an act of connection—through it we become positive, active participants in our lives and in our communities."[1] The performance experience *Shine* is designed to sustain hope through the expressive act of connecting. The script acknowledges our current climate predicament, and proceeds directly to inviting the authorship of solutions, whether in terms of mitigation or adaptation. However, it is not the content of the solutions that gives reason for hope so much as the connections felt in the shared experience of constructive expression. This feeling of connection and the sparking of hope achieved through performance is verified by other writings. In "Devising Green Piece: A Holistic Pedagogy for Artists and Educators," Anne Justine D'Zmura writes, "Good art speaks to the soul and can have long-term impact on those who experience it. The audience members were not the only people changed by the experience. The cast member, designers, crew, and guest experts all recognized the deep and long-lasting results of the experience. The scientists enthusiastically acknowledged the power theatre has to awaken the heart to issues in ways that articles, books, and lectures may not"[2].

That experience of togetherness can be intensified in a performance setting by specifically involving youth performers. Feedback from the tour of *Shine* in New York City and Boulder confirm this. Morana Stipisic shared that, "One thing that I deeply believe in is the fact that the children are the ones who stand a true chance of finding solutions for the

Fig. 4.1   The release of balloons after the final song in *Shine*. Photo by ©2016 Steven Sutton, DUOMO

betterment of our urban future. As Albert Einstein said: 'We cannot solve our problems with the same thinking we used when we created them'."[3] Not only does witnessing youth performance give adults hope, but by engaging youth in a solutions-oriented performance in regards to climate change, we can increase their level of hope and promote positive action. The article, "Motivating Climate Action through Fostering Climate Change Hope and Concern and Avoiding Despair among Adolescents" reveals that by giving youth a feeling that solutions to climate change are within their control, the resulting hope can motivate behavior that benefits other people, their local community, and the world.[4]

Modelling Inclusion

The process by which the Act Two skits are created can serve as a model for how to include all community voices in the sharing of ideas and decision-making. When getting groups together to create the skits for Act Two, mechanisms can be integrated to beckon forth and honor the contributions of all the participating youth. This doesn't mean that every idea expressed by every person will be used as the basis for the skit, but that each voice will be heard and considered. The process I used in New Orleans at Saint Dominic's school in order that each group might arrive at which solution might lend itself best to being acted out for the audience is described in Chap. 3. The process used at the University of East London (Chap. 3) provides a further example. This intentional inclusion can ensure that the group benefits from everyone's perspective. It avoids the repeated pattern of only the most forthright students' voices being heard and their ideas acted upon. Any healthy community needs to benefit from the collected wisdom of the entire group. Indeed, the insights of those within our communities who are perhaps more reserved, who listen and observe before speaking, may be especially valuable, and could be the contributions from which our society currently does not sufficiently benefit. Methods for inclusion from the creative process can be carried over into other types of group work.

Process Over Product

The most valuable outcomes of touring *Shine* are not so much the actual solutions authored by youth, but rather, the youth-led introduction and practice of approaching resilience city planning in an embodied and creative manner. The solutions put forth by youth through the skits in Act

Two represent a beginning engagement in starting to include a more diverse constituency in the planning process. It could be that the splash of a performance project like *Shine* could be conceived as the opening, an announcement, or inauguration of youth participation in resilience planning. It is likely that further, long-term programming is needed to gather more specific input from youth about their self-identified needs, to understand their perspectives, and to gather their contributions.

Five examples of arts-based projects created to include young women's voices in resilience planning for their city are available at Growing Up Boulder (http://www.growingupboulder.org/inside-the-greenhouse.html). Included on this site are proscriptions for leading each of these projects with young women, as well as documentation of the creative artifacts resulting from these activities. These projects were developed as a collaboration between Inside the Greenhouse and Growing Up Boulder and serve as examples of the kind of long-term programming that could be used to encourage youth contributions.

Responsive Content

The South Africa residency of *Shine* reveals that the use of participatory performance to include youth voices in resilience planning benefits from being responsive to the community seeking to include youth voices. As stated by Theatre in Education practitioner David Pammenter, "There is no single way to devise participatory theatre. New creations will take many forms and draw upon many methods and processes. Different cultural contexts will require different kinds of interactions and innovation."[5] In South Africa, the basic play performed was changed to be responsive to the resilience issues identified as primary by the Malope community—stigma around menstruation and early pregnancy—to achieve a more place-based, responsive approach to resilience challenges.

YOUTH CONTRIBUTIONS

Resilience Narratives

It's evident that engaging youth in resiliency planning contributes to their agency, understanding, and self-expression. What they contribute in kind is equally valuable. To begin with, resilience is essentially about change and youth are immeasurably more adept than adults at changing and adapting, and far less attached to set behaviors. Their willingness alone makes their participation in resilience planning essential. Youth also innately bring perspectives that change the narrative. The friendship and scholarship of Bruce Goldstein was both influential and nourishing in the early formation of *Shine*. Not only did he secure me (a Theatre Professor) an invitation to the initial meeting in Boulder with the Rockefeller Foundation 100 Resilient Cities team, his article "Narrating Resilience: Transforming Urban Systems Through Collaborative Storytelling," and our conversations about its contents advanced my understanding of the power of narratives for inclusion of youth voices in the city planning for resilience process. "Change the story, and you change the city."[6] Change who gets to co-author the story, you likewise change the city. Through narrative tales we interpret our condition and identify our most pressing challenges. Narratives enable human actors to respond to these challenges across various ways of knowing while articulating collective identity and "shaping a community of otherwise disparate voices into a coherent and plurivocal vision of the future."[7] A collaboratively can support adaptive behavior. "The essence of resilience is the ability to change as circumstances change, to adapt, and crucially, transform rather than continue to do the same thing faster and better."[8] Performance can be conceived as a rehearsal ground—to hone skills, confidence, and comfort with the ability to change, and, significantly, to change collaboratively with others in community. Putting narrative stories in the service of resilience planning makes sense. We largely act out the stories we tell ourselves about ourselves. This harkens back to noted anthropologist Clifford Geertz's definition of culture as "the ensemble of stories we tell ourselves about ourselves,"[9] and firmly situates the importance of culture in city planning.

Providing an Excuse to Play

Youth are uniquely situated to lead their community in embodied, participatory creative exploration of climate and energy challenges because they know how to play. And they remind adults how to play and try new things, and look at the issues with a fresh perspective. The performance in Boulder at the CU Sustainability, Energy, and Environment Complex (SEEC) was the best example from the tour of youth leading the audience in creative expression. Because this was a six-week residency in the school, there was time to thoroughly work through the youth facilitation of Act Two. With the creation of skits being truly youth-led,

Fig. 4.2 Adults play in the process of rehearsal for *Shine*. Photo by Conner Callahan

there seemed to be a freeing of creativity on the part of the adult participants. At SEEC, when the audience was given streamers to wave during the final number, they easily joined in, seemingly with little self-consciousness (Fig. 4.2). Indeed, since that audience had participated in authoring and performing Act Two, in the final song they were celebrating their own achievements as well as those of the youth in their community.

PERFORMANCE

Increasing Social Capital

Performance can serve as an activating framework accelerating the creation of social capital, which is critical for resilience. In *Building Resilience: Social Capital in Post-disaster Recovery*, author Daniel Aldrich demonstrates that a community's capacity for resilience lies most strongly in the depth of its social capital as evident through robust social networks marked by reciprocity, trust, and cooperation.[10] The *Shine* experience fosters social networking around shared exploration and purpose. For example, although most participants did not know each other before the day of the Boulder NCAR performance, they quickly became a close-knit group that supported each other. Participants made connections, built trust, and helped each other cooperatively towards the common aim of performing a show that contributed to their city's plan for resilience. The generating of social capital through performance was intensified and accelerated by the fact that most of the productions of *Shine* engaged the performers for a full day or a series of days within a week or less. It allowed the arc of understanding to evolve over time and with an intensity of focus. The daily concerns of participants were largely pushed aside as we all had surrendered this time for the preparation of *Shine*. This race-to-the-finish style of performance also fostered the kind of accomplishment felt by participants in a marathon and engendered a heightened feeling of belonging in our city.

Audiences attending *Shine* also contributed to their community's social capital. Since viewing is an action, a viewer of *Shine* actively "observes, selects, compares, interprets. She links what she sees to a host of other things that she has seen on other stages, in other kids of places,"[11] and thereby actively participates in the performance. Performers' actions affect the audience, just as the reactions of the audience affect the actors, generating connections and the "possibility for transformation given in the physical co-presence of actors and spectators. The shared space and time opened up the possibility for transmitting the actors' rhythmic movements onto spectators and transferring them to a state of 'strange intoxication'."[12] Indeed, throughout the tour of *Shine*, there was a feeling of elation shared by everyone—both during and after the performance—based on that reciprocity experienced through performance.

Made for Resilience Planning

One of the goals in writing this book is to understand why, and articulate how, performance as a communicative medium is uniquely well suited to resilience planning. Being embodied, collaborative, and creative, performance can be a highly effective and deeply nuanced tool for exploring as-yet unimagined possibilities for solutions. It provides a forum for witnessing these proposed solutions in real time with real community members. It gives a community the chance to witness possibilities being played out before actual resources are invested and without risk of unintended damage or collateral. It provides the chance to improve upon ideas, try them again, improvise, and to stimulate creative energy and new ways of imagining. By its very cultural heritage, performance brings together community. Performance for youth engagement in city planning is relatively resource efficient, relying largely on community mobilization rather than material resources. It combines the ability to reach many people at once with the personal contact, responsiveness, and potential for dialogue of individual or small group methods. Its biggest draw may be its participatory nature. It achieves the goal of involving a wider constituency in the process of planning and in ownership of the resulting plan.

Dramatic Metaphor

One of the hurdles in addressing climate change and energy use is communicating in ways that inspire and engage people, rather than those that shame, scare, or overwhelm. Addressing these issues through theatre can offer a non-combative approach while taking advantage of the eloquence of metaphor. Dramatic metaphor has the capacity to tease out a more nuanced understanding of both problems and solutions. A movement, action, property, or piece of dialogue can make a comparison to something dissimilar in order to enhance its meaning and to reveal what they might have in common. Metaphor extends beyond one thing merely serving as a symbol for the other. When working with youth, dramatic metaphor can be useful in learning the meaning of complex concepts such as resilience, because it can provide a more visual and active description. For example: the warm-up exercises 'Machine' (described in Chap. 3, Boulder at the NCAR) and 'Resilience in Motion' (used in the University of East London) provide examples of activities that utilize dramatic metaphor to expand the understanding of resilience. This is related to the use of explanatory metaphor to communicate the science of resilience to the public and policymakers that has been studied in order to demonstrate its effectiveness.[13] Since public understanding is key to bridging the research-to-practice divide, utilizing metaphors and bringing them to life through performance can increase the understanding of the concept of resilience.

The power of a metaphor in performance can be drawn from the connection to the raw material used in constructing the comparison. For example, the dramatic metaphor of weaving together a community was especially powerful in the Navajo Nation where there is a deep cultural investment in weaving. Associations such as the warmth of a grandmother's blanket, sheering sheep, gathering plants to dye the threads, or the beauty of a shawl on a young girl's shoulders in a ceremony, all get mixed into the experience of the metaphor. These associations link "weaving" with the forming of community, such that the common feelings of warmth, vibrancy, and pride contribute to a deeper understanding and experience of both. According to CU Environmental Design Professor Bruce Goldstein, co-author of the article "Narrating Resilience: Transforming Urban Systems Through Collaborative Storytelling," resilient communities need a foundational narrative that "envelopes concerns of those affected and weaves these concerns into a credible story that resolves issues and ties people together."[14]

James Geary, author of the book *I Is an Other: The Secret Life of Metaphor and How It Shapes the Way We See the World,* asserts that metaphor "systematically disorganizes the common sense of things—jumbling together the abstract with the concrete, the physical with the psychological, the like with the unlike—and reorganizes it into uncommon combinations."[15] He goes on to say that metaphor, "shapes our view of the world, and is essential to how we communicate, learn, discover, and invent, yet we typically fail to recognize it" as it "takes place mostly outside our conscious awareness."[16] This is powerfully relevant in *Shine,* where the use of metaphor seeks to jumble the many factors at play to invigorate youth towards fresh ways of approaching both global and local challenges. Putting metaphor into performance means that a sideways scattering of the hips distributes the harvest seeds. The heartbeat of a drum pumps life into ancient animals. A pole thrust into the earth signals ownership of the black gold beneath its depths. A storm of fossil fuel flags disrupts the natural carbon cycle, its wreckage ripping through the fabric of community. With arms reaching towards the sky, youth stretch for new solutions with outstretched hands. The use of metaphor here reaches into the bones and muscles; it needn't be intellectually understood to be effective, in fact, it's more effective if it is first felt, sung, danced, and experienced. Invention can arise from unlikely configurations and improvised action.

The Importance of Celebration

It is a tradition in cultures throughout the world to end public gatherings with an inspirational song and dance that expresses the achievement of the gathering's purpose. A song combined with synchronized movement can nudge the follow-through from concern to action regarding the purpose addressed at the gathering. If you leave humming the final tune, you carry the spirit of commitment with you into your daily life. It can infuse your thoughts and actions with the inspiration that was built into the event purposefully by its organizers. Participating in song can provide an experience of connection and joy, which allows the singer to feel the value of community that in turn will hopefully strengthen the resolve to act on its behalf.

Shared cultural expression unites us, allows us to feel who we are as a community and provides a medium to communicate that beyond our borders. The final song may even set the attitude for moving forward towards new behavior. Attitudes inform behavior. Without the right attitude, certain behaviors don't make sense; if you don't have an attitude of generosity, sharing your lunch doesn't make sense. If we are hoping adults are willing change behavior, we best attend to our own attitude.

The culminating song is best characterized as being accessible, catchy, and affirmational for group membership. In her book *Utopia in Performance*, Jill Dolan "investigates the potential of different kinds of performance to inspire moments in which audiences feel themselves allied with each other, and with a broader, more capacious sense of a public, in which social discourse articulates the possible, rather than the insurmountable obstacles to human potential."[17] When people dance and wave streamers in rhythm to a beat, they literally take up more room as they expand into the public realm, and, despite their many differences, feel part of a unified beat, pumping life through their shared community. Songs and anthems are a vital tool in nation building, and can be used for city spirit building as well.

Dissolving Boundaries

At its heart, theatrical performance is just a human agreement, an agreed upon time and space around which we erect a wall of some sort, within which we allow greater levels of imagination, wondering, and questioning. We expect and reward with thunderous applause more expansive behavior. We allow and even expect disruption of the status quo. Without disruption theatre lacks the spark to burn. But when the performance within this wall of theatre is participatory and applied to real-life issues of consequence—such as a city's plan for resilience—what happens? The wall delineating life from performance can disintegrate underfoot as community members step across into new roles, imagining a better world and coming together to create it. Then life itself can become more expansive, disruptive, and full of wonder, movement, and joy.

CONCLUSION

Climate change is scary. Change and uncertainty is scary. Our current communications—our words, graphs, and furrowed brows—are not producing the level of engagement needed to address our challenges and imagine solutions that allow each and every one of us to thrive. So, it's time to bring in the experts: the ones who remember how to change, who are as nimble, bright, and full of energy and ideas as humans get. The ones who don't yet drive, so aren't attached to big cars. Who don't know grass doesn't grow on rooftops, or that a car can't be run on water as fuel. When they are the ones writing the policies—sooner than we think—we'll be grateful we included them in imagining the cities and world we are creating. Youth need to be included now, and we need their energy, insight, and audacity.

This project was primarily designed for cities taking part in the 100 Resilient Cities Initiative because these cities are actively creating a plan or strategy for resilience. All communities though—whether included in the original 100 or not—would benefit from the use of performance for activating youth to engage in the authorship of their future resilience. After all, "theatre remains the only place where the audience confronts itself as a collective."[18] This performance can be used in any community to kick-start a wider community conversation. By nature of the fact that it is most often mounted in schools, the performance alerts the community that their city is undergoing a resilience planning process, which is often the first time citizens are aware of such efforts. And as any community organizer knows, if you want adults to get interested, involve their kids. For these reasons, many city planners would welcome the contributions of youth to their plans but lack methods for conversing with youth or receiving their input. It is hoped that if this performance is implemented as part of a larger community effort, that sustained support of youth in resilience planning may result. It is my hope that the example of this performance will inspire other creative offerings that can involve youth through performance towards envisioning the cities we want.

City policies and plans often get put on a shelf and collect more dust than results. What is needed is community buy-in so that real individuals integrate those policies and plans into action in their daily lives, and the life of their community. Since performance as an expressive tool is rooted in action, it seems uniquely well suited for planning resilience action. "Drama means action. Theatre is the place where an action is taken to its conclusion by bodies in motion in front of living bodies that are to be mobilized."[19] Being embodied, collaborative, and creative, performance can be a highly effective tool for exploring yet unimagined possibilities. It provides a forum for witnessing proposed solutions in real time with real community members who can be inspired to mobilize them in their daily lives. It gives a community the chance to witness possibilities being played out before actual resources are invested and with minimal risk of unintended damage or consequences. It provides the chance to improve upon ideas, try them again, improvise, and to stimulate creative energy and new ways of imagining. It infuses joy and creativity into the entire process. It could be that the inclusion of joy is possibly the most sustaining ingredient in ensuring continued engagement by a larger constituency. We'll come back to something time and again if it makes us feel good. Performance offers a highly time-efficient, cost-effective, nuanced, and fun approach to resilience planning. Fun could be one of the most sustainable yet largely untapped forces for galvanizing communities into action on behalf of their community's resilience.

Youth are often identified as being disruptive. If there was ever a time to disrupt the narrative of the energy and climate, it's now. *Shine* invites and celebrates youth's disruption of the status quo. Tom Wasinger, the composer of *Shine*, lovingly describes the rehearsal process as "controlled chaos." The freedom of thought, preposterous ideas, radical concern, and outright silliness that youth have brought to city planning through the tour of *Shine* has been exceptional. This theatrical approach offers a viable alternative mode for exploring, thinking, and creating modes for living in this world. In his final chapter in the book *Readings in Performance and Ecology*, Wallace Heim concludes with, "Every skill and faculty of inventiveness is needed in a time of environmental change, including the delight in making rude combinations, the pleasure of rigorous scholarship, and the rewards of fearless experimentation."[20] What we do in these coming decades will determine if we can thrive or even survive on this planet as a species. *How* we plan for our future and *who* we include in the planning may determine what that future looks like. Participatory performance by youth is one way to shine a light on a brighter future. See http://www.insidethegreenhouse.org/shine for more tools.

NOTES

1. Michael Rohd, *Theatre for Community Conflict and Dialogue: The Hope Is Vital Training Manual* (Portsmouth, NH: Heinemann, 1998), xix.
2. Anne Justine D'Zmura, "Devising Green Piece: A Holistic Pedagogy for Artists and Educators," in *Readings in Performance and Ecology* (Palgrave Macmillan, 2012), 179.
3. Morana Stipisic, "Performance at The City We Need," October 28, 2016.
4. Kathryn Stevenson and Nils Peterson, "Motivating Climate Action through Fostering Climate Change Hope and Concern and Avoiding Despair among Adolescents," *Sustainability* 8, no. 1 (2016): 1–10.
5. David Pammenter, "Theatre as Education and a Resource of Hope: Reflections on the Devising of Participatory Theatre," in *Learning Through Theatre: The Changing Face of Theatre in Education*, 3rd ed. (New York: Routledge, 2013), 83.
6. Bruce Goldstein et al., "Narrating Resilience: Transforming Urban Systems Through Collaborative Storytelling," *Urban Studies* 52, no. 7 (May 2015): 1290.
7. Ibid., 1300.
8. Ibid., 1287.
9. Clifford Geertz, *The Interpretation of Cultures: Selected Essays*, First Edition (New York: Basic Books, 1973), 448.
10. Daniel Aldrich, *Building Resilience: Social Capital in Post-Disaster Recovery* (The University of Chicago Press, 2012).
11. Jacques Raciere, *The Emancipated Spectator* (Brooklyn, NY: Verso, 2009), 13.
12. Erika Fischer-Lichte, *The Transformative Power of Performance: A New Aesthetics*, trans. Saskya Iris Jain (New York City: Routledge, 2008), 193.
13. Nathaniel Kendall-Taylor and Abigail Haydon, "Using Metaphor to Translate Science of Resilience and Developmental Outcomes," *Public Understanding of Science* 25, no. 5 (2016): 576–587.

14. Goldstein et al., "Narrating Resilience: Transforming Urban Systems Through Collaborative Storytelling," 1290.
15. James Geary, *I Is an Other: The Secret Life of Metaphor and How It Shapes the Way We See the World* (New York: Harpers Perennial, 2012), 3.
16. Ibid.
17. Jill Dolan, *Utopia in Performance: Finding Hope at the Theater* (University of Michigan Press, 2005), 164.
18. Raciere, *The Emancipated Spectator*, 5.
19. Ibid., 3.
20. Wendy Stein and Theresa May, eds., *Readings in Performance and Ecology* (New York, NY: Palgrave Macmillan, 2012), 216.

References

Aldrich, Daniel. *Building Resilience: Social Capital in Post-Disaster Recovery*. The University of Chicago Press, 2012.

Dolan, Jill. *Utopia in Performance: Finding Hope at the Theater*. University of Michigan Press, 2005.

D'Zmura, Anne Justine. "Devising Green Piece: A Holistic Pedagogy for Artists and Educators." In *Readings in Performance and Ecology*, 169–80. Palgrave Macmillan, 2012.

Fischer-Lichte, Erika. *The Transformative Power of Performance: A New Aesthetics*. Translated by Saskya Iris Jain. New York City: Routledge, 2008.

Geary, James. *I Is an Other: The Secret Life of Metaphor and How It Shapes the Way We See the World*. New York: Harpers Perennial, 2012.

Geertz, Clifford. *The Interpretation of Cultures: Selected Essays*. First Edition. New York: Basic Books, 1973.

Goldstein, Bruce, Anne Wessells, Raul Lejano, and William Hale Butler. "Narrating Resilience: Transforming Urban Systems Through Collaborative Storytelling." *Urban Studies* 52, no. 7 (May 2015): 1285–1303.

Kendall-Taylor, Nathaniel, and Abigail Haydon. "Using Metaphor to Translate Science of Resilience and Developmental Outcomes." *Public Understanding of Science* 25, no. 5 (2016): 576–587.

Pammenter, David. "Theatre as Education and a Resource of Hope: Reflections on the Devising of Participatory Theatre." In *Learning Through Theatre: The Changing Face of Theatre in Education*, 3rd ed., 83–102. New York: Routledge, 2013.

Raciere, Jacques. *The Emancipated Spectator*. Brooklyn, NY: Verso, 2009.

Rohd, Michael. *Theatre for Community Conflict and Dialogue: The Hope Is Vital Training Manual*. Portsmouth, NH: Heinemann, 1998.

Stein, Wendy, and Theresa May, eds. *Readings in Performance and Ecology*. New York, NY: Palgrave Macmillan, 2012.

Stevenson, Kathryn, and Nils Peterson. "Motivating Climate Action through Fostering Climate Change Hope and Concern and Avoiding Despair among Adolescents." *Sustainability* 8, no. 1 (2016): 1–10.

Stipisic, Morana. "Performance at The City We Need," October 28, 2016.

Appendix

Choreography Worksheet with Counts and Lyrics for Each Dance in *Shine* with Open Spaces for Choreographer Notes

Long Time Comin'

2 counts of 8 Introduction/Instrumental

4 counts of 8 Vocals

> *Long time comin', a long time a comin' comin', long time comin' along*
> *Long time comin', a long time a comin' comin', long time comin' along*
> *Long time comin', a long time a comin' comin', long time comin' along*
> *Long time comin', a long time a comin' comin', long time comin' along*

6 counts of 8, fades out

Harvest

4 counts of 8 Instrumental for Harvesters

8 counts of 8 Vocals for Harvesters

> *Harvesters: We plant together standing side by side,*
> *We reap together with our arms open wide.*
> *We work together with the seeds that we sow*
> *We feed each other with the foods that we grow (twice)*

© The Editor(s) (if applicable) and The Author(s) 2017
B. Osnes, *Performance for Resilience*,
DOI 10.1007/978-3-319-67289-2

4 counts of 8 Instrumental for Fossil Fuels

4 counts of 8 Vocals for Fossil Fuels

Foss: We don't tend sheep anymore
We don't harvest wheat anymore
Sister don't be such a bore
We get our food from a store
We don't sleep at night anymore
Cause it ain't such a fright any more.
Sister don't be such a bore
Get out and dance on the floor

8 counts of 8 Vocals for Sol, Foss, Seed Sower

Sol: You must agree that your ways are wasteful
You must agree that my path is more tasteful

Foss: I bought the shoes on my feet
Drive my car down the street
It's hard to believe that we're related
Sis, your ways are antiquated

Seed Sower: It's the harvest party, let's not fight
Or waste our time on who's wrong or right
Brother, sister now let's get along
And weave together the words of these songs

8 counts of 8 Mash Up Between Harvesters and Fossil Fuels Singing Simultaneously, Fades Out

Harvesters: We plant together standing side by side,
We reap together with our arms open wide.
We work together with the seeds that we sow
We feed each other with the foods that we grow (2xs)
(and)
Foss: We don't tend sheep anymore
We don't harvest wheat anymore
Sister don't be such a bore

We get our food from a store
We don't sleep at night anymore
Cause it ain't such a fright any more.
Sister don't be such a bore
Get out and dance on the floor (2xs)

Weaving Song

(Note: Counts are not provided for this song since the task of weaving is more of a movement pattern than a rhythmic dance.)

Over thread and under strand
Over time we understand
Fibers will combine to be
The fabric of community

Ancestry and history
Cloth to warm us, hold and form us

Sun is constant always there
Rays of light weave through the air
Come on out and sow the seeds
Simply we can meet our needs
Ancestry and history
Cloth to warm us, hold and form us

Over thread and under strand
Over time we understand
Fibers will combine to be
The fabric of community

Progress Song

1 count of 8 Instrumental for Fossil Fuels

6 counts of 8 Vocals (after which music degrades, switches into sounds of storm)

Foss: Come with me to fuel the world I'm looking for coal
Come with me to fuel the world I'm looking for oil

Come with me to fuel the world I'm looking for gas
Fuel to meet increasing needs to move fast.
Just you and me, I'm energy.
Just you and me, I'm energy.

Finale/Shine

4 counts of 8 Instrumental for Entire Cast

15 counts of 8 Vocals for Entire Cast

Turn around touch the ground
'Til a new idea is found
Look up, look down, shake up your town
Swish your feet, repeat
Right down our main street
Light bright feels right
Run for fun in the sun and
Shine shine shine shine
Shine shine shine
Shine shine shine

Turn around touch the ground
'Til a new idea is found
Look up, look down, shake up your town
Swish your feet, repeat
Right down our main street
Light bright feels right
Run for fun in the sun and
Shine shine shine shine
Shine shine shine
Shine shine shine

INDEX

A

Active culture, 5
Agriculture, 44
Ally Agreement, 69
Animal agriculture, 100
Applied Theatre, 6

B

Balloons, 97, 99, 109, 111
Berger, Nicole, 18
Biomass, 9
Boal, Augusto, 98, 106
Boulder, CO, 62, 69, 70, 72, 74–76, 78–82, 86, 87, 89, 90, 103, 116, 126, 132
Breed, Ananda, 94

C

Carbon credits, 128
Carbon cycle, 50, 56
Carboniferous Period, 7, 26, 41, 72, 125
Carter, Crista, 88
Casey Middle School, 78–80, 86

Catholic Church, 104, 105, 113
Celebration, 145
Choreographer. *See* Arthur Fredric and Lisa Denton
Cities, 4
  city planning, 7
Climate change, 5, 67, 80, 87, 98, 146
Climate communication, 111. *See also* Conference on Communication and Environment
Community-Based Adaptation to Climate Change, 6
Composer. *See* Wasinger, Tom
Compost, 83
Conference on Communication and Environment, 74
Connecticut, USA, 122, 123
Cornell, Eric, 80, 84
Corpos Christi, 105
Cross-curricular project week, 97

D

Dance, 86, 88
Denton, Lisa, 35, 36, 123, 124

© The Editor(s) (if applicable) and The Author(s) 2017
B. Osnes, *Performance for Resilience*,
DOI 10.1007/978-3-319-67289-2

Diné Bikéyah. *See* Navajo Nation
Dramatic Metaphor, 143, 144

E
EcoAmerica, 111
Ecocriticism, 6
Ecodramaturgy, 6
Embodiment, 14
Environmental stewardship, 107, 111
Evolution, 107
Extinction, 125

F
Fabric of community, 11, 47, 144
Flood, 72, 75, 96, 101
Forgotten People, 69
Fossil Fuels, 8, 12, 42, 44, 46, 66, 73, 105, 129
Fredric, Arthur, 3, 34–37, 39, 88, 90, 92, 122

G
Gender, 62, 116, 131
GeneSister, 79, 85
Goldstein, Bruce, 140, 144, 148
Growing Up Boulder, 79, 139
Guibert, Greg, 74, 76

H
Habitat, United Nations, 88, 90
Hackett, Claire, 114–118
Harvest, 9, 36
Homelessness, 96
Hope, 85, 136, 138
Hopi, 63, 66
Hunter Elementary School, 89–91

I
Image Theatre, 96
Inclusion, 138
Industrial Revolution, 47, 98, 103, 125, 127
Inside the Greenhouse, 5, 78
Inter-disciplinary. *See* Cross-curricular project week
Intergovernmental Panel on Climate Change, 2
International Environmental Communication Association, 74

K
KenCairn, Brett, 87

L
Lanko, Paty Romero, 2, 70, 86
Limpopo Province, South Africa. *See* Malope, South Africa
London, UK, 94, 95, 98, 99, 101

M
Malope, South Africa, 64, 67, 114, 118, 122
Manhattanville College, 122, 124
Manygoats, Adrian, 63, 64, 69
Menstruation, reducing stigma, 114, 116, 117
Metaphor. *See* Dramatic metaphor
Moench, Marus, 80
MUSE Studio, 79

N
National Center for Atmospheric Research, 2, 69, 71, 127

National Dance Institute, 124
National Renewable Energy
    Laboratory, 3
Navajo Nation, 63, 64, 88, 144
Navajo Women's Energy Project, 64,
    68
New Orleans, LA, 104, 109, 138
New School, NY, 88
New York City, NY, 64, 88, 94, 124,
    126, 133, 134
*New York Convening on the City We
    Need*, 88

O

On Care for Our Common Home, 104,
    113, 134

P

Peace Corps, 114
Performance Studies, 6
Photosynthesis, 2, 8, 42, 73
Play, 85, 86, 140
Pollution, 66, 96, 101, 110
Pope Francis, 104, 111, 132
Pregnancy, teen, 114, 115, 119, 121
Privilege, 69
Process, 42, 138

R

Racism, 101
Raderstorf, Joellen, 80
Ramsing, Soren, 94, 97–99, 105, 134
Resilience, 4, 91, 94, 95
Resilience narratives, 139
Responsive, 121, 139

Riverside Primary School, East
    London, 93–95, 97, 98, 102
Rockefeller Foundation 100 Resilient
    Cities Initiative, 3, 140

S

Saint Dominic's School, 104, 105,
    112, 115, 138
Sandy Hook, CT, 124
Science Discovery Camp, 125, 126,
    129
Science, Technology, Engineering, and
    Math (STEM), 125
Science, Technology, Engineering,
    Arts, and Mathematics (STEAM),
    129
Skit, 81
Social and Environmental Transition-
    International (ISET), 80
Social capital, 141, 142
Solar, 76, 79, 82, 91, 110, 125
Sommer, Shelly, 87
Sperling, Joshua, 3, 19, 37, 38, 71,
    90, 124
Stay-cation, 81
Stipisic, Monica, 92, 136
Sun, 7, 67
Sustainability, Energy, and
    Environment Complex CU, 140

T

Theatre history, 112
Traffic, 100
Triassic Period, 9, 125
Trilobite, 72
Tuba City, AZ, 62–64, 94, 103, 115

U

United Nations Habitat. *See* Habitat, United Nations

University of East London, 93, 94, 138

Urban, 7, 37, 45

Urban Thinkers Campus, 88, 90

V

Values, 120

Vandever, Gerald, 63, 65, 67

Veganism, 100, 129

W

Wasinger, Tom, 3, 31, 77, 92, 124

Weaving, 32, 67, 144

White, James, 3, 78, 80

Wind, 91

Y

Young Women's Resilient Voices, 79

Youth, 7, 147, 148

youth-led, 76

CPSIA information can be obtained
at www.ICGtesting.com
Printed in the USA
LVHW08*0219230818
587881LV00018B/206/P